The missing half

The missing half
Girls and science education

edited by
ALISON KELLY

Those who have never entered upon scientific pursuits know not a tithe of the poetry with which they are surrounded.

Herbert Spencer

Manchester University Press

Published by
Manchester University Press
Oxford Road, Manchester M13 9PL

British Library Cataloguing in Publication Data

The missing half.
1. Science- Study and teaching (Secondary) – Great Britain
2. Education of women – Great Britain
I. Kelly, Alison
507'.12'41 Q183.4.G7

ISBN 0–7190–0753–4 (paper)
ISBN 0–7190–0831–X (cased)

Computerised Phototypesetting
by G.C. Typeset Ltd., Bolton, Greater Manchester

Printed in Great Britain
by Butler & Tanner Ltd
Frome and London

CONTENTS

ACKNOWLEDGEMENTS

I would like to express my thanks to all the people who have helped to make this book possible. In the background are the unknown children, parents, teachers and schools who provided the data for research and contributed their own experiences and insights. Then come the friends, colleagues and families who help every author to work out her or his ideas and theories. Over the years I have discussed the problems of girls and science with many people, some of whom may recognize their ideas in this book. If your point has become so much a part of my thinking that I no longer remember its origin, please take this as a compliment and accept my thanks. More easily identifiable are those who have helped in the later stages of production: Linda Ollerenshaw, Jean Ashton and Sue Hardy who typed and retyped with great good humour; Mrs Watts who drew the diagrams; Judy Samuel who read and commented on the completed typescript from a teacher's point of view; and Ray Offord of Manchester University Press who took trouble with details of style and spelling. And last but not least the contributors, who accepted deadlines, delays and cuts with remarkably little protest.

THE CONTRIBUTORS

Judy Bradley took a BA and MA in sociology at Essex University and a DPhil in Education at Oxford University. She has carried out research at Oxford into pre-school education and into science education in the secondary sector. She is now Senior Research Officer at the National Foundation for Educational Research, where she is researching into further education teaching as a career.

John Collings graduated from Manchester University in 1968 with a BSc (Hons) degree in psychology and then read for an MSc in Clinical Psychology at Leeds University. Since taking up a lecturing post he has become increasingly interested in psychological factors involved in subject choice and took his PhD in this field at Bradford University in 1978. Since taking his first degree he has conducted research into the psychophysiology of mental illness, taught psychology to medical, biological science, nursing and social work students, and tutored for the Open University and Leeds University's Extra-Mural Department. He is at present Senior Lecturer in psychology at Leeds Polytechnic, where he teaches general and applied psychology to social science students.

Dave Ebbutt taught science in Nigeria and Uganda before becoming Head of Science in a Fenland secondary school. He is now Research Fellow at the Cambridge Institute of Education, working on the SSRC Cambridge Accountability Project.

Maurice Galton is a Senior Lecturer in Education at Leicester University, where he lectures on the Psychology courses in Research Methodology and Classroom Interaction. He taught chemistry at an all-boys school and then at Leeds University before coming to Leicester as a researcher with the Schools Council Project for the Evaluation of Science Teaching Methods. He is now co-director of a five-year programme of research examining the effectiveness of different teaching styles in junior and middle schools.

Jeffrey Gray took a BA in Modern Languages and then a second one in Psychology, both at Oxford University. He trained in Clinical Psychology and did a PhD in Experinental Psychology, both at the Institute of Psychiatry, University of London. He is at present a lecturer in the Department of Experimental Psychology, Oxford University, and a Fellow of University College, Oxford.

Dorothy Griffiths is a Lecturer in Sociology at Imperial College of Science and Technology, University of London, where she specialises in teaching courses on Science, Technology and Society to scientists and engineers. With Esther Saraga she is working on a project examining the ways in which biological theories have been used to explain and justify the subordination of women; her other interests include career structures in science. She is active in the Women's Liberation Movement, and is a member of the Editorial Collective of *Feminist Review*; she has also worked in the British Society for Social Responsibility in Science for a number of years.

Jan Harding taught chemistry and physics for some years in both mixed and girls' schools. She spent time at home when her two children were small, before working in a college of education. She was seconded for two years to the Curriculum Diffusion Research Project, based at Chelsea College, which enquired into the use of the Nuffield Science Teaching Project, and later directed the work of the Girls and Science Education Project which examined boys' and girls' performance in 1974 O level science examinations. Since 1975 she has held the post of Head of Chemistry Section at the Centre for Science Education, Chelsea College.

Alison Kelly took a BSc in Physics and a MSc in Astrophysics, and then taught maths and science for two years in Swaziland. During this time she became interested in the science education of girls, and subsequently pursued this interest for a PhD in Education. She is now a lecturer in Sociology at Manchester University, where she teaches courses on Women in Society, Sociology of Education and Research Methods. With Judith Whyte she is currently engaged in an intervention project to devise strategies for use in schools to persuade more girls to study science. She is also a member of the Women and Education newsletter group.

Elinor Kelly took a BA degree in Sociology, then spent a year teaching English in a girls' secondary school in Northern Sudan before returning to London to study for an MPhil in Social Anthropology. Subsequently she carried out a research study of a youth work project in Manchester, and, since 1976, has been teaching Applied Social Studies in Manchester University's Extra-Mural Department. Her special interests include Third World women and racial minorities in Britain.

Peggy Newton has a BA and an MA in Psychology and is currently working towards a PhD in Social Psychology. She is a Lecturer in Social Psychology at Huddersfield Polytechnic where she teaches courses on Sex, Gender and Society, Human Development and Research Methods. Her research on girl technicians in engineering is continuing with the assistance of Professor G. M. Stephenson and the Social Psychology Research Unit at the University of Kent at Canterbury.

Milton Ormerod graduated from Manchester University with Honours in Chemistry in 1942. He taught chemistry for twenty years before becoming a lecturer in chemistry method in the Education Department of Brunel University. He has developed extensive elementary and advanced 'learning packages' for teaching chemical structure and its relations to properties. At

Brunel he has supervised several research students and gained a PhD for research on the measurement of attitudes to science and school subject preferences, teacher liking and subject choice and their interrelationships. He now holds the position of senior lecturer and research tutor at Brunel University.

Judy Samuel gained her BSc in Biochemistry and then took a Postgraduate Certificate in Education. She is now teaching chemistry in a comprehensive school, and is a member of the Manchester based Women and Education newsletter group.

Esther Saraga is a Senior Lecturer in Psychology at the Polytechnic of North London. As a psychologist her interests include women's role in the family, and psychological sex differences, and she is particularly concerned to examine biological and psychological theories in the context of women's subordination. She is active in the Women's Liberation Movement and on the Editorial Collective of *Feminist Review*.

Alan Smithers took a first in Botany and a PhD in Plant Physiology and became a lecturer in Botany at Birkbeck College, London. There he got very interested in Education and moved to a Research Fellowship at Bradford University. He retrained as a psychologist and obtained an MSc in the Psychology and Sociology of Education and a PhD in Education and has been elected ABPsS. He has written a book on sandwich courses, three research reports and over fifty articles on various aspects of education. His interest in girls in science goes back to his student days and why were there so few! He is currently Professor of Education at the University of Manchester.

Helen Weinreich-Haste describes herself as a developmental-social psychologist. Her research interests are in adolescent moral development, and sex roles. Her work on the latter has led her increasingly into looking at the educational implications of the image of science. She lectures in Psychology at the University of Bath, and has been involved for several years in Women's Studies courses.

Judith Whyte took an MA degree in Philosophy and English Literature at Glasgow University and a postgraduate Certificate of Education at Oxford University. She taught in schools in Glasgow and London before becoming Educational Organiser for Oxfam for two years. Since then she has worked as a researcher and part-time tutor on Women's Studies courses in adult education. She is currently lecturing in Humanities at Manchester Polytechnic, where she teaches a fourth year BEd Honours option on Sex and Gender in the Classroom. With Alison Kelly she is now working on an Affirmative Action project aimed at improving schoolgirls' attitudes and achievement in science.

INTRODUCTION

1

Girls and science education: is there a problem?

ALISON KELLY

In 1906 W. Felter argued that girls should not be taught physical science except at the most elementary level, because the expenditure of nervous energy involved in the mastery of analytic concepts would be injurous to their health. Such an argument may seem ludicrous today, but how far have we really progressed? Even now far fewer girls than boys study physical science. Is the present-day rationale for this any more tenable than Felter's? And what, if anything, should be done about it?

This volume presents theoretical essays, research studies and personal accounts which attempt to answer these questions by focusing on the science education of girls. Collectively, the essays are concerned to elucidate the origins of girls' under-achievement in science and suggest ways to improve their performance. The contributions concentrate on the secondary education of girls in England. This is because it seems likely that improvements to girls' science education will be most readily effected in secondary schools. Girls' under-achievement in science is a complex phenomenon, with many of the characteristics of a vicious circle linking socialisation, schooling, the job market and family responsibilities. But the circle can be broken into, and it is probable that this can be done most effectively at the secondary school stage. Most girls encounter formal science education for the first (and last) time in secondary school. If they drop out of science at this stage they are unlikely ever to take it up again. So success at subsequent stages is dependent upon success in secondary science. Although the foundations for girls' attitudes and abilities in science are probably laid before they reach secondary school, it is difficult to intervene in the home or the primary school. Few parents or primary school teachers are especially concerned with science, and few are aware that they may be influencing their children's later achievement in science. This makes it difficulty to reach them or persuade them to modify their behaviour. Similarly it is difficult to intervene at the professional level. Employers will not take the problems of women in science seriously until women constitute a substantial proportion of the scientifically trained labour force and demand to be taken seriously. And this will not happen until large numbers of girls are educated to a high level in science at school.

The prospects for change in secondary schools are more hopeful. Many secondary science teachers are aware and concerned that girls are not succeeding as well as boys in science. They are also confused and puzzled over the causes and remedies for this situation. This pool of concerned teachers is potentially a powerful force for change and improvement in girls' science education. New ideas can spread relatively rapidly through professional contacts such as subject associations, HM Inspectors and in-service courses. At present most teacher training courses make little if any reference to sex differences or sexism in the classroom. But the initial training of science teachers could easily be modified to include material on sex differences in science achievement and suggestions as to how to encourage girls. It is therefore to the audience of concerned science teachers and teacher trainers that this book is chiefly addressed, in the hope that, by combining analysis of the current state of girls' science education with suggestions for action based on that research, it will help to crystallise their concern and encourage them to take practical steps to improve the situation.

The present situation in England and Wales[1]

Is there in fact a problem with girls and science education?[2] Let us start with some facts and figures. The Department of Education and Science survey carried out in 1973 (DES, 1975) found that fewer girls than boys were offered the chance of studying physics, and to a lesser extent chemistry, in their fourth and fifth years at secondary school, but that more girls than boys were offered the opportunity to study biology (Table 1.1). The sex differences were, however, considerably more marked in choosing and taking these subjects than in being offered them. Thus almost half of all fourth- and fifth-year boys studied physics, compared to only 12 per cent of the girls. Conversely nearly half of all fourth- and fifth-year girls studied biology, compared to only 28 per cent of the boys. These differences were quite accurately reflected in the proportions of 1975–76 school leavers of each sex who had attempted CSE or O level in each subject. The figures for O level passes also show boys predominating in physics and chemistry but girls predominating in biology. Most pupils of both sexes attempted a certificate in maths, but more boys than girls passed maths O level.

These figures immediately indicate two points which will recur throughout this book. The first is that biological and physical sciences must be distinguished in any discussion of girls' science education. Physics and chemistry are definitely 'boys' subjects' in the sense that many more boys than girls study them at school. But the reverse is true for biology, which, at least up to the fifth form, is studied by more girls than boys. The second point indicated by Table 1.1 is that informal barriers against girls studying physical

Table 1.1 The percentage of girls and boys being offered, taking and obtaining qualifications in science subjects in the last years of compulsory schooling

	(a) Being offered (% of fourth and fifth form pupils)	*(b)* Choosing (% of col. (a))	*(c)* Taking (% of fourth and fifth form pupils)	*(d)* Attempting CSE or O level (% of 1975–76 school leavers)	*(e)* Passing O level* (% of 1975–76 school leavers)
Physics					
Girls	71	17	12	9	5
Boys	90	52	47	40	16
Chemistry					
Girls	76	22	17	12	6
Boys	79	35	27	25	12
Biology					
Girls	95	52	49	46	18
Boys	88	31	28	27	12
Maths					
Girls	n.d.	n.d.	n.d.	68	20
Boys	n.d.	n.d.	n.d.	70	27

Note. n.d. = no data.

* Grades A-C and CSE Grade 1.

Source. (*a*), (*b*), (*c*), Department of Education and Science, *Curricular Differences for Boys and Girls*, Education Survey 21, table 6, HMSO, 1975; (*d*), (*e*) Department of Education and Science, *Statistics of Education, 1976*, Vol. 2, table 8, HMSO, 1977.

science are more important than formal barriers. Over 70 per cent of girls had the formal opportunity to study physics in their fourth and fifth year, but only 17 per cent of the girls who were offered this opportunity actually took it. With the introduction (since this survey was carried out) of the Sex Discrimination Act, the proportion of girls being formally offered the chance to study physics has probably increased still further. But legislation will not help much if girls lack the motivation to continue with the physical sciences, or if they are informally discouraged from so doing. The ways in which informal barriers can operate against girls in science are an important subject for discussion in this book.

The ratio of men to women studying pure science subjects at various levels of the education system is shown in Table 1.2 (a). It is clear that the proportion of women is generally lower at higher levels of qualifications. This trend is most noticeable if the main academic track from O and A level to university is considered. Four times as many boys as girls attempt O level physics, but seventeen times as many men as women obtain doctorates in physics. Even in biology, which is taken mainly by girls at O level, there are equal numbers of men and women by A level, and four times as many men as women gaining PhDs. Chemistry falls between physics and biology, and mathematics is similar to chemistry. Thus at higher academic levels all the pure sciences are dominated by men. CSE is usually taken by school leavers, rather than as a preliminary to advanced study in a subject, but the high ratio of males to females doing CSE physics is noticeable. This is probably because it is a useful qualification for technical apprenticeships, which are taken far more often by boys than by girls (see Table 1.4 below). Similarly, advanced courses in further education are both more technically oriented and less likely to lead on to further study than first degrees at university, and the ratio of males to females is correspondingly higher.

The bottom line of Table 1.2 (a) shows that the upper reaches of the educational system are dominated by men in all subjects, not just in science. This being so, it is reasonable to ask whether girls are specifically under-represented in science, or whether their position in science merely reflects their position in the educational system generally. This is examined in Table 1.2 (b), which shows the over- or under-representation of females in each branch of science, relative to their representation in education as a whole. A value greater than 1 indicates that there is a higher ratio of men to women in that subject than in education as a whole at that level, i.e. women are less well represented in that subject than in education as a whole. The

Table 1.2 The representation of males and females in pure science education in England, 1975–76

					Universities		
	Attempt CSE	*Attempt O level*	*Attempt A level*	*Advanced FE courses*	*Obtain first degree*	*Do post graduate research*	*Obtain PhD*
(a) *The number of males for every female studying science subjects at various levels of the educational system*							
Biology	0·4	0·6	1·1	2·0	1·3	2·6	3·8
Mathematics	1·0	1·5	3·8	3·8	2·6	8·0	9·7
Chemistry	2·1	2·1	2·4	6·7	4·7	7·0	9·7
Physics	7·5	3·8	4·9	13·1	7·2	9·6	17·1
All subjects	1·1	1·0	1·4	3·1	2·0	4·1	6·3
(b) *The representation of females in science compared to their representation in the educational system as a whole at various levels*[a]							
Biology	0·4	0·6	0·8	0·6	0·7	0·6	0·6
Mathematics	0·9	1·5	2·8	1·2	1·3	2·0	1·5
Chemistry	1·9	2·0	1·8	2·2	2·4	1·7	1·5
Physics	6·8	3·7	3·5	4·2	3·7	2·4	2·7

Notes

a The figures in Table 1.2(b) are: 'number of males for every female studying subject' divided by 'number of males for every female studying all subjects' at that level'.

(i) At A level, mathematics includes all mathematical subjects; at university level, biology includes all biological subjects, but excludes medical and paramedical students.

(ii) The university figures refer to all British universities.

(iii) The figure for 'all subjects' shows more boys than girls attempting CSEs. This is because there are more boys than girls in the population; it does not indicate that girls are under-achieving at this level.

Source. Statistics of Education, 1976, Vol. 2, and *1975*, Vols. 3 and 6.

result is striking. The increasing dominance of men at succeeding levels of the educational system which was evident in the raw data of Table 1.2(a) has disappeared. In Table 1.2(b) each subject has a characteristic ratio which remains fairly constant at the various stages of the educational system. In other words, from O level onwards, girls drop out of science at approximately the same rate as they drop out of other subjects. Girls are relatively over-represented in biology, under-represented in mathematics and chemistry, and very under-represented in physics. This under- or over-representation is evident at O level, and persists through to doctoral level. But women's position in science, *relative to other subjects*, does not get any worse at higher levels of the educational system.

It is clear from this analysis of the statistical data that two sets of variables are operating. The factors which cause girls to drop out of education between O level and PhD operate in science as in other subjects and account for the sharp increase in the ratio of males to females found in Table 1.2(a). But the factors which are specific to girls dropping science appear to operate before O level. This point is crucial in the context of remedial action. Measures to encourage girls to study science will probably be most effective if applied at a stage prior to O levels; measures to keep girls in the educational system as a whole should operate at and beyond A levels.

Trends in girls' representation in the pure sciences can be assessed by comparing the situation in 1975–76 (Table 1.2) with that in 1965–66 (Table 1.3). There has been some improvement in girls' position in the educational system as a whole in the intervening ten years. The ratios of males to females in all subjects are consistently smaller in 1975–76 (Table 1.2(a)) than in 1965–66 (Table 1.3(a)). The ratios of males to females studying science subjects are also smaller in 1975–76 than in 1965–66 (except for the predominantly female subject of biology at CSE and O level, where the proportion of boys has increased). But the relative under-representation of girls in science has hardly altered (although it does appear to have stabilised across all levels of education, possibly owing to the changing structure of the educational system: compare Table 1.2(b) and Table 1.3(b)). In 1965–66 girls studied science, particularly physics, less frequently than they studied other subjects at all stages of education. The situation in 1975–76 was similar. Girls' position in science had improved slightly, but this improvement mirrored the improvement in girls' position in the educational system as a whole. The characteristic under-achievement in science remained. Taking a rather longer time span

Table 1.3 The representation of males and females in pure science education in England, 1965–66

	Attempt CSE	Attempt O level	Attempt A level	Advanced FE courses	Universities: Obtain first degree	Universities: Do post graduate course[a]	Universities: Obtain PhD
(a) *The number of males for every female studying science subjects at various levels of the educational system.*							
Biology	0·4	0·5	1·2	2·0	1·4	3·8	7·0
Mathematics	1·5	1·9	5·7	4·8	2·9	8·2	28·0[b]
Chemistry	5·5	3·2	3·8	11·8	6·3	16·1	19·8
Physics	16·4	5·2	6·0	10·7	7·7	18·4	32·0
All subjects	1·4	1·2	1·9	8·1	2·6	3·9	11·8
(b) *The representation of females in science compared to their representation in the educational system as a whole at various levels*[c]							
Biology	0·3	0·3	0·6	0·2	0·5	1·0	0·6
Mathematics	1·1	1·6	3·0	0·6	1·1	2·1	2·4[b]
Chemistry	3·9	2·7	2·0	1·5	2·4	4·1	1·7
Physics	11·7	4·3	3·2	1·3	3·0	4·7	2·7

Notes

a In the 1965–66 statistics no distinction was made between postgraduate research and other postgraduate courses.

b Figures for 1966–67. The figures for 1965–66 are anomalous because only one woman (compared to 142 men) obtained a doctorate in mathematics in that year. When the number of women obtaining doctorates is so small the PhD ratios are liable to pronounced fluctuations from year to year.

c The figures in Table 1.3(b) are: 'number of males for every female studying subject' divided by 'number of males for every female studying all subjects at that level'.

(i) At O and A level mathematics includes all mathematical subjects; at A level biology includes botany and zoology; at university level biology includes all biology subjects, but excludes medical and paramedical subjects.

(ii) The university figures refer to all British universities.

Source. Statistics of Education, 1966, Vols. 2, 3 and 6.

shows that the percentage of graduates in science subjects who were women fell between about 1930 and 1950 (Kelly, 1979). Only recently has the percentage been as high as it was in the 1920s (although the absolute numbers are, of course, much larger). Medicine is an exception to this pattern, since the proportion of women doctors qualifying has risen steadily since 1920 (despite the imposition, until recently, of a quota on the proportion of women admitted to medical schools); technology is also an exception, with the proportion of women being almost constant around 0·5 per cent from 1920 to 1960, but rising in the last fifteen years to almost 5 per cent.

A certain level of education in the natural science is generally a prerequisite for study or employment in the applied sciences. For example, there are no large entry O or A levels in medicine or engineering, and university entrants in these subjects are expected to have qualifications in pure science. For this reason most of the studies in this volume concentrate on the natural sciences, which are seen as more fundamental. However, education in the applied sciences is also widespread, and, of course, it is vocationally linked. Table 1.4 examines the representation of girls and women in this sphere. It is not usual (although it is possible) to progress through the applied sciences from one stage to the next, and so the different parts of Table 1.4 do not form a sequence (as do most of the columns of Tables 1.2 and 1.3). Nevertheless it is interesting to compare women's representation at different stages.

The ratio of males to females attempting CSE and GCE examinations in applied science is shown in Table 1.4(a). Although girls dominate domestic science almost as strongly as boys dominate technical drawing, applied sciences and metalwork, this equivalence is more apparent than real. The scientific content of the subjects may be comparable, but, as Byrne (1978) points out, their vocational content is not. Whereas 'boys' crafts' are useful qualifications for technical apprenticeships, 'girls' crafts' are only relevant to the housewife role. Even aspiring chefs are better advised to take biology and chemistry than domestic science.

Turning to further and higher education (Table 1.4(b)), there are several interesting features. Although men predominate in engineering and technical subjects at all levels, this is more marked in further education than at university. Possibly university is seen as less deviant than further education for able girls following an unusual course. Or perhaps employers are reluctant to sponsor girls on further education courses in engineering. Men constitute a higher

Table 1.4 The representation of males and females in applied science education in England, 1975–76.

(a) *The number of males for every female attempting CSE or GCE qualifications in applied science*

	Attempt CSE	*Attempt O level*	*Attempt A level*
Technical drawing	54·5	53·7	84·8
Applied sciences[a]	46·2	62·2	–
Metalwork	134·6	253·9	254·5
Domestic science/cookery	0·1	0·03	0·005

(b) *The number of males for every female studying applied science in further or higher education, 1974–75*

	Further education		*Universities*		
	Non-advanced	*Advanced*	*Obtain first degree*	*Post graduate study*	*Obtain PhD or MD*
Engineering and technical subjects	53·4	49·0	22·8	19·5	44·3
Paramedical subjects	0·1	0·8	1·2	2·0	3·3
Medicine	–	–	2·4	2·2	2·9

(c) *The number of male and female apprentices and trainees in scientific fields, 1971 (10% sample).*

	M	*F*	*M : F ratio*
Craft level:			
Electrical and electronic	6,921	54	128·2
Engineering and allied trades	23,163	120	193·0
Semi-professional and technician level:			
Draughtsmen	1,547	112	13·8
Laboratory assistants	954	560	1·7
Technical and related	1,766	123	14·4
Engineers	1,512	n.d.	n.d.
Nurses	676	5,772	0·1
Total	63,639	14,182	4·5

Notes

a The subjects included under applied sciences are building and engineering studies at CSE and engineering workshop theory and practice, design and technology, and building studies at O level. There are no large-entry subjects of this type available at A level.

n.d. = no data, i.e. the number of females was negligible and so not included in the tables.

Source. Statistics of Education, 1976, Vol. 2, 1975, Vols. 6 and 3. *1971 Census, Economic Activity*, Vol. 2, table 10.

proportion of those obtaining PhDs than of those obtaining first degrees at university, but in this respect engineering is similar to other subjects (cf. Table 1.2). Paramedical subjects show a strong concentration of women in non-advanced further education (studying subjects such as nursing and occupational therapy), but an increasing concentration of men in advanced further education and at university (studying subjects such as pharmacy and pharmacology). In medicine itself there are two to three times as many male as female students.

Among apprentices and trainees in science-related fields the proportion of women is again extremely high in nursing. In fact nurses account for over 40 per cent of all female apprentices and trainees (service trades, mainly hairdressing, account for another 26 per cent). The greatest concentration of male apprentices and trainees is in engineering and allied trades (36 per cent) and in electrical and electronic trades (11 per cent), but there are very few women in these two areas. Indeed, with the exception of engineering, women are better represented at semi-professional and technician level than at craft level. Women laboratory assistants are common, and although there is a large majority of male trainee draughtsmen (*sic*) the proportion of women among trainee draughtspeople (draughters?) is much larger than the proportion of women among students attempting technical drawing at CSE or O or A level. This suggests that the schools are not keeping up with the demand from girls for skills in this area.

These statistics confirm many commonsense impressions of applied science. Although women are well represented in medical science, they are found in the occupational roles which are supportive rather than innovatory and which do not require advanced study. Women are also well represented among laboratory assistants, another supportive job. Men predominate in engineering at all levels. Thus when it comes to applied science, the sex biases found in pure science (girls to biology, boys to physical science) are exaggerated.

The present situation around the world

Is the under-representation of women in physical science unique to England? I have investigated this using data from the UNESCO *Statistical Yearbook* to compute the percentage of tertiary level science students who were women, in eighty countries (Kelly, 1976a). These countries were then grouped into six geographical regions. Table

1.5(a) gives the average number of men for every woman student in each region.

In all six regions women formed less than half the tertiary level students of natural science. However, the range was enormous, from

Table 1.5 The representation of women among science students around the world

(a) *The average number of men for every woman among science students in six regions of the world*

	Number of men for every woman in:			
	Natural science	*Engineering*	*Medicine*	*All tertiary education*
African (13)	9·5	70·4	5·7	6·8
Arab (11)	3·4	27·6	2·6	3·4
West European (22)	3·0	24·6	2·2	2·0
Asian (15)	2·5	22·3	1·9	2·6
Central and South American (14)	1·2	16·2	1·4	1·9
East European (9)	1·0	3·5	0·7	1·4

(b) *The representation of women among science students compared to their representation in tertiary education as a whole in six regions of the world*[a]

	Natural science	*Engineering*	*Medicine*
African (13)	1·4	10·4	0·8
Arab (11)	1·0	8·1	0·8
West European (22)	1·5	12·3	1·1
Asian (15)	1·0	8·6	0·7
Central and South American (14)	0·6	8·5	0·7
East European (9)	0·7	2·5	0·5

Notes

[a] The figures in Table 1.5(b) are: 'number of men for every woman in discipline' divided by 'number of men for every woman in all tertiary education'.

(i) The number of countries in each region is given in brackets after the name of each region.

(ii) The data refers to the years 1969, 1970 or 1971.

(iii) 'West European' refers to countries with a Western type of culture. It includes Israel, Japan and Australasia.

Source. Kelly, 1976a (derived from UNESCO *Statistical Yearbooks*)

one in nine in Africa to 49 per cent in Eastern Europe. Comparison of individual countries showed an even greater range; women were under 5 per cent of science students in Saudi Arabia, Zaire, the Congo and Togo, but over 60 per cent in the Philippines, Poland, Paraguay, Guatemala and Bulgaria (Kelly, 1976a). Thus, although women's under-representation in science is widespread, it is by no means universal.

As in the English situation, it is reasonable to examine women's relative under-representation in science compared to other subjects. This can be done for the international figures by comparing the number of men for every woman in tertiary education as a whole with the number of men for every woman in science. Table 1.5(a) also shows that women formed considerably less than half the total student body in all six regions of the world. But only in the African and West European regions were women under-represented in natural science relative to other subjects (indices greater than 1 in Table 1.5(b)). Thus the situation in England is typical of Western countries, but is not replicated in other regions of the world. The UNESCO *Statistical Yearbooks* do not distinguish biology, chemistry and physics students, but figures for the science-related disciplines of engineering and medicine are given, and these are also summarised in Table 1.5. It is clear that women are drastically under-represented in engineering, although less so in Eastern Europe than elsewhere. Women also form less than half the medical students in most regions of the world, but only in West European countries are they under-represented in medicine relative to tertiary education as a whole.

In a rather different type of international study (Kelly, 1978), I used data from the IEA survey of Science Education in Nineteen Countries (Comber and Keeves, 1973) to study sex differences in science achievement among fourteen-year-old pupils. The results of this study are discussed in detail in Chapter 2. The main findings were that in all the countries studied (mainly Western but including Hungary and Japan) boys achieved better than girls on the science tests, and the gap between the sexes was approximately constant. Nevertheless girls in some countries achieved better than boys in other countries. This demonstrates that in some circumstances girls can achieve well in science. In all the countries studied the sex difference varied in the same manner between the different branches of science, being negligible in biology and very pronounced in physics, with chemistry intermediate.

Taken together, these international studies demonstrate several things. Girls' under-achievement in science is by no means confined to England. It is a common phenomenon, particularly but not solely in the Western world. But international studies also show that girls can do well in science. In some countries women form a large proportion of science students, and in most regions of the world they are as well represented in science as in the rest of education. Girls in some countries achieved better in science than boys in other countries. Nevertheless within each country studied girls achieved worse than boys. And in all the countries the sex differences were more pronounced in physical sciences than in biological sciences.

The problem

This, then, is the intellectual problem: why do girls, relative to boys, under-achieve in science? Why is this under-achievement consistently evident in so many different countries? And why is it so much more marked in physics than in biology, with chemistry intermediate?

But the problem is not solely an intellectual one. There is also a practical problem for all who want to encourage girls in science. There are several perspectives on this practical aspect. The most common approach concentrates on society's need for scientists and technologists, and bemoans the loss of able girls from the supply of scientific manpower (*sic*). But there is also a more radical, feminist approach, concerned with social inequality. Girls who cannot or will not learn science are cut off at an early age from a wide range of careers and interests. By conforming to a feminine stereotype which excludes science they are moving towards traditional women's occupations, and the low pay and low status which frequently accompany such occupations. Girls who succeed in science have a wider choice than those who fail, so the feminist seeks to reduce failure.

The feminist position has implications not only for individual women but also for women as a social group. Women should have access to professions which depend on scientific training, since members of these professions control many aspects of life in an industrial society. Science is changing society, and these changes should not be directed exclusively by men. So substantial numbers of women should become professional scientists and technologists. And all women should have sufficient scientific understanding to enable them to take a full and informed part in current debates on such issues as environmental pollution, energy resources and genetic engineering. At a practical level science is important for a feeling of competence and environmental mastery. A woman in an industrial society is surrounded by machinery and labour-saving devices; often she does not understand or control this equipment sufficiently even to carry out simple repairs. She is encouraged to rely on 'experts', almost invariably men. This encourages feelings of incompetence and inadequacy. Women become passive consumers of an incomprehensible male-structured environment. More adequate scientific education for girls could help to combat this situation.

For educators too, girls' under-achievement in science is a practical problem. Science is widely considered to be an essential part of general education, without which a person's experience is incomplete. In Hirst's (1969) philosophy of education, science comprises one of the seven fundamentally distinct ways of perceiving the world to which every complete individual should have access. A society which allows and encourages girls to opt out of scientific experiences is thereby denying them a part of their intellectual heritage as human beings. Scientists who themselves find science a meaningful and worthwhile activity are probably especially conscious of this; they are often keen to introduce pupils to a system of thought and way of viewing the world which they themselves find stimulating and rewarding. But present-day arrangements are failing many girls in this respect. The question is, why? And what can be done about it?

Contributions to the book

These are the questions which this book is centrally concerned to answer: why do girls, on average, achieve less well than boys in science? And how can girls' achievement in science be improved?

The book is divided into three sections. The first part presents a variety of theoretical positions on girls' under-achievement in science. Some of these positions are complementary, others are competing. In order to make this section more coherent the authors were all asked to relate their theories to the same set of empirical results. These results are set out in Chapter 2 where I describe in more detail my international study of sex differences in the science achievement of fourteen-year-olds. Several explanatory hypotheses embodying commonsense reasoning on girls' under-achievement in science were tested in that study. But these all turned out to be unsatisfactory, at least in the crude form in which they were first set out. The following chapters offer alternative explanations. In

Chapter 3 Jeffrey Gray suggests that biological differences between males and females may account for the difference between boys' and girls' achievement in science. The other authors in this section all prefer social explanations but they differ in emphasis. Elinor Kelly argues in Chapter 4 that in patriarchal societies (which means most, if not all, societies today), girls and boys are socialised to different roles by pressure from parents, teachers and peers. In the following chapter I suggest that children are essentially self-socialising, and that once girls have defined themselves as feminine and science as masculine they will tend to avoid it. In Chapter 6 Esther Saraga and Dorothy Griffiths take a more structural view. They argue that science today is a hierarchical, competitive activity, which reflects the nature of capitalist society. Until society and science are reformed, girls and women will always be excluded.

This theoretical section is intended to give some idea of the different ways in which girls' under-achievement in science can be conceptualised. It provides a framework—or, more accurately, a variety of possible frameworks—within which the more empirical studies can be located. The second section consists of accounts of research on aspects of subject choice in school and the factors which may encourage or discourage girls in science. In contrast to section one, which uses cross-national comparisons, section two is mainly concerned with England. Sex differences in science achievement are an international phenomenon, and as such they require a theoretical explanation which is not confined to any one country. But education systems vary widely between countries, and if we hope to implement changes through the schools it is important to use the existing framework in England as a starting point. This is not to deny that comparative works can be valuable. We might learn much from a study of girls' science education in the Soviet Union, for example, if such a study existed. But more often educational strategies are tied to particular educational systems. Accounts of grade-skipping pre-calculus classes in the United States (Fox, 1976) have little direct application to the English situation (although they may indicate strategies which can be adapted for use in Britain).

The first two chapters in the second section are both concerned with the way choices between subjects are made, but they differ in their approach. In Chapter 7 Milton Ormerod describes several large-scale projects on the relationships between attitudes to science and liking and choosing science. In contrast to this, Dave Ebbutt uses a case study of one school in Chapter 8 to explore the reasons given by girls and their teachers for choosing or dropping science in the fourth year. In Chapter 9 I too consider the reasons pupils give for choosing or dropping science, this time using a survey approach. But my main concern in the chapter is to document the way in which early choices between school subjects structure later career and educational opportunities.

The focus of the next three chapters is rather different: they all look at groups of girls who defy convention and choose to specialise in science. Peggy Newton interviewed girls who were training to be engineering technicians, and in Chapter 10 she discusses some of the factors which have helped or hindered these girls in their unusual decision. In Chapter 11 Judy Bradley describes a longitudinal study which shows that future science specialists can be distinguished from future arts specialists as early as the third year of secondary school on the basis of intelligence and personality measures. In Chapter 12 Alan Smithers and John Collings take a closer look at girls studying science in the sixth form, and explore their personality and motivation through psychometric scales and in-depth interviews. Happily the results confirm, on the whole, the conclusions of the previous chapter concerning the difference between science and arts specialists.

Chapter 13 is one of the few contributions which considers what actually goes on in science lessons, although the approach is that of an outsider rather than a participant. Maurice Galton looks at the different teaching styles employed by science teachers and assesses the exposure and reactions of girls and boys to these varying styles. In Chapter 14 Jan Harding compares the achievement of girls and boys in science examinations. She pays particular attention to the differences between single-sex and co-educational schools, between Nuffield and conventional examinations and between different types of examination question.

The last two chapters in the research section are concerned with the image of science. Dave Ebbutt asked second year pupils in a mixed comprehensive school to identify bits of the science syllabus as 'for boys' or 'for girls'. In Chapter 15 he discusses the results and explores the criteria which differentiate 'boys' science' from 'girls' science'. Helen Weinreich-Haste used rating scales and sentence completion techniques to elicit the image of science held by third and fourth year school children. In Chapter 16 she describes this image and discusses the accuracy and possible implications of the stereotypes.

Theoretical arguments and research studies can seem very remote from what goes on in classrooms. The third section of the book comes closer to the everyday experiences of pupils and teachers with a series of personal accounts of science lessons. Many of the more theoretical points are echoed here, but it is also clear that much of science education consists of interpersonal encounters, and that individual teachers can have a strong influence on the attitudes and achievement of girls in science. In Chapter 17 schoolgirls (both present and past) describe their science lessons, their feelings about science, and the factors that influenced their choice of options. In Chapter 18 the teachers have their say: Judy Samuel writes in some detail about her experience as a feminist teaching chemistry in a mixed comprehensive school, and other teachers have contributed shorter accounts of the differences they perceive between girls and boys in science, and the methods they use to encourage girls in science. Finally, in Chapter 19, Judith Whyte suggests how teacher training institutions could sensitise their students to sexism in schools, and makes some immediate suggestions for practising science teachers.

The contributors were all asked to draw out the practical implications of their findings and to suggest what, if anything, should be done on the basis of these findings to improve the position of girls in science. In the final chapter I try to pull these suggestions together and provide a more comprehensive account of possible ways to improve girls' achievement in science. The task of summing up is made more difficult by the diversity of political and ideological positions represented in the book. Most of the women authors are feminists and active in the Women's Liberation Movement. Although this shared involvement gives them much common ground, it does not imply a uniformity of views; feminists can be reformist or revolutionary, integrationist or separatist. The non-feminist authors also vary in their approach. Inevitably, then, the final chapter reflects my own perspective rather than the distilled wisdom of all the contributors. When pushed, I define myself as a liberal feminist. As a liberal, I believe in presenting all sides of a question; as a feminist I am concerned to improve the position of women. The composition of the book echoes this approach by presenting a variety of viewpoints on one crucial determinant of women's position in society.

The contributors have tried as far as possible to avoid the use of technical terms. They all describe their findings in everyday language. However, some statistical terms were unavoidable if the results of the research studies were to be described adequately. These terms are explained in an appendix, which is intended to assist readers who are unfamiliar with statistics. Other readers may find that there is insufficient detail in some of the chapters. Several of the studies have already been reported in academic journals or books, and reference is made to these for further details where appropriate.

This book does not attempt to duplicate the systematic research reviews provided by Gardner (1975) and Kelly (1976b). Nor does it offer a complete coverage of the subject of Women and Science, but focusses more narrowly on the educational issues. The problems faced by women making their careers in science are not treated, although there is now a considerable body of research on this topic (see for example Bayer and Astin, 1975; White 1970). Nor is there any discussion of the way science affects the lives of women—an under-researched but crucial area (see Brighton Women and Science Collective, 1980).[3]

Notes

1 For brevity, and with apologies to Welsh nationalists, England and Wales will be referred to as England in the remainder of this chapter. Unless otherwise stated the information does not apply to Scotland (which has a separate education system) or to Northern Ireland.

2 This and the following section draw heavily on material already published by Kelly and Weinreich-Haste (1979).

3 I would like to thank Judy Samuel and Elinor Kelly for reading and commenting on an earlier draft of this chapter.

References

Bayer, A. E. and Astin, H. S. (1975) 'Sex differentials in the academic reward structure', *Science*, Vol. 188, 796.

Brighton Women and Science Collective (1980) *Alice through the Microscope; the Power of Science over Women's Lives*, Virago.

Byrne, E. M. (1978) *Women and Education*, Tavistock.

Comber, L. C., and Keeves, J.P. (1973) *Science Education in Nineteen Countries*, Almqvist & Wiksell.

Department of Education and Science (1975) *Curricular Differences for Boys and Girls*, Education Survey, 21, HMSO.

Department of Education and Science (1977), *Statistics of Education 1976*, Vol. 2, HMSO.

Felter, W. L. (1906) 'The education of women', *Education Review* Vol. 31, 351.

Fox, L. H. (1976) 'Sex differences in mathematical precocity: bridging the gap' in Keating, D. P. (ed), *Intellectual Talent: Research and Development*, Johns Hopkins University Press.

Gardner, P. L. (1975) 'Sex differences in achievement, attitudes and personality of science students: a review', *Science Education: Research* 1974.

Hirst, P. H. (1969) 'The logic of the curriculum', *J. Curriculum Studies*, Vol. 1, 142.

Kelly, A. (1976a) 'Women in physics and physics education' in Lewis, J. (ed), *New Trends in Physics Teaching*, Vol. III, UNESCO.

—(1976b) 'Women in science: a bibliographic review', *Durham Research Review*, Vol. 7, 1092.

—(1978) *Girls and Science: An International Study of Sex Differences in School Science Achievement* Almqvist & Wiksell.

—(1979) 'Where have all the women gone?' *Physics Bulletin* Vol. 30, 108.

Kelly, A., and Weinreich-Haste, H. E. (1979) 'Science is for girls?', *Women's Studies International Quarterly*, Vol. 2, 275.

White, M. S. (1970) 'Psychological and social barriers to women in science', *Science*, Vol. 170, 413.

THEORETICAL INTERPRETATIONS

2

Sex differences in science achievement: some results and hypotheses[1]

ALISON KELLY

A striking result emerged from the IEA science survey (Comber and Keeves, 1973).[2] On average, boys scored considerably better than girls in science achievement tests in nineteen countries. Thanks to a grant from the Spencer Foundation, I was able to undertake a research project with IEA to investigate the sex differences in science achievement further.[3] The project is reported in full elsewhere (Kelly, 1978). In this chapter I will summarise the main results and outline the explanatory framework within which I tried to account for these results.

Knowing that, overall, boys had achieved better than girls on the IEA science tests, my research was chiefly concerned with exploring the reasons for this sex difference. I began the study with three general hypotheses, centring respectively on cultural, school and attitudinal factors. Simply stated, the culture hypothesis suggests that girls do not do as well in science as boys because of the social expectation that they will under-achieve; the school hypothesis suggests that science is taught in schools in a way more suited to boys than to girls; and the attitude hypothesis suggests that girls achieve less well than boys in science because they have less favourable attitudes towards science. These hypotheses represent the most straightforward, non-biological explanations of sex differences in science achievement, and in various forms they are frequently proposed. Although none of them turned out to be completely satisfactory, they will be discussed here in an attempt to decide which aspects of these popular explanations can be retained and which must be rejected. In this respect negative results can be as important as positive results. Each hypothesis will be evaluated in turn, and the evidence bearing on it assessed. Biological explanations were not considered in my study because there was no information on biological factors available in the IEA survey.

Data

The IEA science data have been fully described by Comber and Keeves (1973), so I will only give brief details here. Standardised science tests were administered (after translation) to nationally representative samples of between 500 and 7,000 fourteen-year-olds in fourteen developed countries (Australia, Belgium (Flemish- and French-speaking), England (and Wales), Finland, Germany (Federal Republic), Hungary, Italy, the Netherlands, New Zealand, Scotland, Sweden and the United States). Four developing countries were also studied, but in most of these countries the tests proved too difficult for most of the pupils and very low average scores were recorded. For this reason I omitted the developing countries from my analyses. Ten-year-olds and pre-university pupils were also tested in most countries, but I chose to concentrate on the fourteen-year-old age group because most pupils had been exposed to some formal science teaching by this age, and yet the results were not complicated by school drop-out.

The science tests were compiled by an international committee from items submitted by each country and pre-tested in at least six countries. An elaborate screening and selecting procedure was used to avoid bias towards the science curriculum of any one country, and, in addition, teachers provided estimates of the proportion of their pupils who had had the chance to learn the material involved in each item. The final tests are reprinted in Comber and Keeves (1973, pp. 330–64). They are of the multiple choice type and are divided into biology, chemistry, physics and practical sub-tests according to content. The practical sub-test is mainly concerned with laboratory practice, covering such areas as which arrangement of apparatus or which experimental procedure is most suitable for a particular problem.

In addition to the science achievement test, pupils completed a word knowledge test, consisting of synonyms and antonyms (also reprinted in Comber and Keeves, 1973, pp. 393–4). Although clearly neither culture-free nor tapping the whole range of academic skills, this test did provide an independent measure of ability against which to compare the pupils' performance in science. Background information on the pupils' homes and schools was obtained from questionnaires completed by the pupils and their teachers, and several attitude scales were also included in the test battery (these are reprinted in Peaker, 1975). The main testing took place in 1970.

The culture hypothesis

The most widely advocated and intuitively reasonable hypothesis of the origin of sex differences in science achievement is the culture hypothesis. The culture hypothesis suggests that girls achieve less well than boys in science because society does not encourage or expect girls to achieve as well as boys in science. In its most general form the mechanism for this hypothesis consists of an assertion that, throughout the world in present-day society, achievement by women is neither expected, nor encouraged, nor adequately recognised when it occurs. The image of women presented by the media and in children's literature is a passive, submissive one. Women occupy the supportive and expressive roles in society, men the innovative and instrumental ones. Women are respected for vicarious achievement, through their husbands and children, but direct personal achievement is considered hard and unfeminine. In these circumstances girls reduce their aspirations and fulfil society's expectations of them by under-achieving. This general proposition is superficially attractive and may provide a basis for understanding women's place in adult society. But it is unsufficiently explanatory for the present purposes; at fourteen years old girls do not under-achieve in all spheres, but only in certain fields. Girls achieved better than boys in the IEA tests of literature, reading and languages, and within science boys' advantage was much larger in physics and practical than in chemistry and biology (Walker, 1976) A mechanism which refers only to socialisation with respect to a generalised achievement motivation cannot explain this varied pattern of sex differences.

The operation of the culture hypothesis can, however, be refined so as to focus on the specifically masculine connotations of science. In this form it suggests that girls are socialised away from science at an early age by virtue of the toys they are given to play with, the hobbies they are encouraged in, the household jobs they are asked to help with, and the masculine image of science and scientists in books, films and television. These factors operate more strongly in physical science than in biological science, which may be considered nurturant and therefore suitable for girls. Languages and literature, with their liberal arts and cultural connotations, are also suitable for young ladies, hence, so the argument goes, girls' superior achievement in these subjects.

This view of the culture hypothesis carries with it the implication that, where the socialisation of girls with respect to science varies, so too does girls' science achievement. There were no specific data on socialisation processes in the IEA survey, so a direct test of the hypothesis was not possible. The test had instead to take the form of assuming that the expectations for girls in science differ in different countries, and looking for concomitant variation in the sex differences in science achievement. It must be emphasised that this is only an assumption. It can be argued intuitively that the fourteen countries studied differ in their conception of woman's place, attitudes ranging from the Scandinavian emphasis on equality to the Latin *mama* and *machismo* and the Anglo-Saxon 'feminine mystique'. But it is not clear how real these differences are. For the purposes of the argument it is not necessary to suggest that any of the national cultures are non-sexist with respect to science—only that some are more sexist than others. We can then examine whether these assumed differences in sex role aculturation are reflected in differences in science achievement.

The main results of the international study are presented in Fig. 2.1. Within each country fourteen-year-old boys gained higher scores than fourteen-year-old girls in each branch of science. This despite the fact that in some countries science was compulsory, and therefore both boys and girls had studied the same amount of science up to the age of fourteen. The only exception to boys' superior achievement was in chemistry in Belgium (Flemish), and there were problems with the sampling in Belgium (see Comber and Keeves, 1973, p. 48). These national sex differences are the fundamental phenomena considered in this study.

However, international comparisons are also valid, and these show that girls in some countries performed better than boys in other countries. Figure 2.1 also illustrates this point. Hungarian girls did better in biology than boys in all other countries, and Japanese girls did better in biology than boys in all countries except Hungary and Japan. More generally there was almost complete overlap between the distribution of country mean scores for girls and for boys in biology. The separation was slightly greater in chemistry, but still Hungarian and Japanese girls achieved better mean scores than boys in all other countries. In physics there was much less overlap—the mean score for girls in twelve countries was lower than the lowest mean score for boys in any country. Nevertheless Japanese girls did achieve a mean score in physics which was comparable to that of boys in other countries. This single result is of crucial importance because it illustrates that, with a suitable mixture of cultural background, attitude, motivation and teaching,

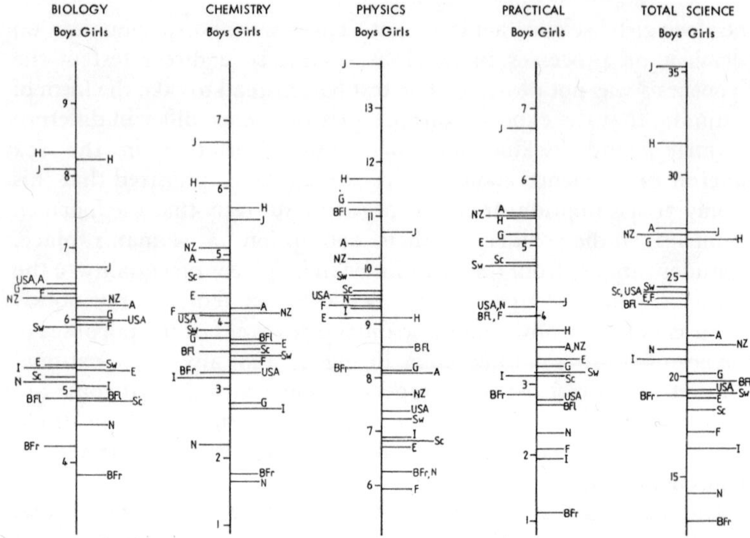

Fig. 2.1 Mean science achievement scores in four branches of science.
Notes
(a) A – Australia; BFI – Belgium (Flemish); BFr – Belgium (French); E – England; F – Finland; G – Germany; H – Hungary; I – Italy; J – Japan; N – Netherlands; NZ – New Zealand; Sc – Scotland; Sw – Sweden; USA – United States.
(b) The scale for each subject is proportional to the average within-country standard deviation of scores. This enables between-subject comparisons of dispersion to be made by eye.

the mean score of a group of girls can exceed the boys' international average score in physics. The composition of this mixture of factors is another question. The practical sub-test resembles physics in exhibiting a large separation between boys' and girls' mean scores.

The same information is given in the form of standardised sex differences (SSD) in Table 2.1.[4] As already noted, the difference between the sexes was fairly small in biology (averaging 13 per cent of a standard deviation),[5] slightly larger in chemistry (averaging 21 per cent) and very marked in physics and practical (averaging 61 and 49 per cent respectively).[6] Combining all four branches, boys achieved better than girls by an average of 48 per cent of a standard deviation in total science.[7] An alternative, and perhaps more immediately comprehensible, way of expressing this difference is that 68 per cent of boys reached or exceeded the score achieved by 50 per cent of the girls in total science. The corresponding figures for biology, chemistry, physics and practical were 55 per cent, 58 per cent, 73 per cent and 69 per cent respectively of the boys reaching the standard set by 50 per cent of the girls.

A remarkable feature of Table 2.1 is the uniformity of the sex differences in any one subject area. In nearly all countries the SSD were small in biology and large in physics, with chemistry and practical intermediate. No one country stood out from the others as exhibiting peculiarly large or peculiarly small differentiation between girls and boys in all branches of science. Indeed, the sex differences in science achievement were far more characteristic of a particular branch of science than of a particular country. The overall level of achievement varied considerably between countries, with pupils in some countries achieving far better than pupils in other countries. But the gap between the sexes remained fairly constant whatever the overall level of achievement. This suggests that the same factors improved and depressed performance for girls and boys. The international range of combinations of cultural, home and school influences produced an international spread of mean scores. But altering the combination affected both sexes similarly, and all

Table 2.1. The standardised sex differences in science achievement

	Biology	*Chemistry*	*Physics*	*Practical*	*Total Science*
Australia	8	17	52	51	40
Belgium (Flemish)	0	–5	64	43	40
Belgium (French)	34	54	53	61	71
England	2	17	51	41	34
Finland	14	20	82	61	61
Germany (FRG)	13	29	69	55	56
Hungary	3	10	54	45	37
Italy	9	16	56	42	43
Japan	34	29	61	60	55
Netherlands	20	19	81	60	71
New Zealand	1	25	57	51	42
Scotland	9	25	57	42	41
Sweden	19	11	61	44	45
United States	15	24	52	40	40
Mean	13	21	61	49	48

Note. Positive standardised sex differences indicate that boys achieved better scores than girls.

combinations surveyed here produced better science achievements with boys than with girls.

These results do not strengthen the culture hypothesis. Although the societies considered here ranged from those overtly committed to equality of the sexes to those with much more traditional attitudes, the standardised sex differences in science achievement were similar in all the developed countries. Overall achievement (presumably determined by a combination of societal and pedagogical influences) varied from country to country, but the SSD (which the cultural hypothesis implies were determined by the same influences) remained almost constant across countries. The most straightforward interpretation of this finding is that the sex differences in science achievement were relatively impervious to cultural influences. Factors peculiar to a branch of science seemed to be of greater importance in producing sex differences than factors characteristic of the national culture or school system.

An alternative interpretation of the near constancy of the sex differences is that the specific cultural factors which affected sex differences in science achievement were uniform across the countries considered. It is clearly difficult if not impossible to quantify the relevant factors. But it is striking to notice that the proportion of science students who were women, which ranged from 54 per cent in Hungary to 13 per cent in Japan, bore no relation to how well girls did in science in each country. This result suggests that social and cultural factors have more influence on girls' decisions whether or not to continue to study science than on their level of achievement.

Within each country the ratio of boys to girls increased as higher levels of achievement were considered. But internationally the gap between boys and girls was slightly narrower in countries where the mean level of science achievement was higher. Thus there was no absolute level beyond which girls' performance fell off relative to boys'; the fall-off occurred at a different level in each country, and was relative to the boys in that country.

In twelve countries data were available from the IEA survey on both ten- and fourteen-year-old pupils. Standardised sex differences in science achievement were computed for ten-year-olds in the same way as for fourteen-year-olds. The SSD for ten-year-olds were considerably smaller than those for fourteen-year-olds, but showed many of the same patterns. In particular, the SSD for ten-year-olds were again larger in physics than in chemistry or biology, and more characteristic of a subject area than of a country. Nevertheless there was a strong correlation ($r = 0{\cdot}70$) between the SSD for a country's ten- and fourteen-year-olds. This may indicate that sex differences in achievement at fourteen were built upon foundations already laid at ten. In countries where the sex difference was relatively large at ten years old it tended to be large also at fourteen years, and conversely where the sex difference was small at ten years old it tended to be small also at fourteen years old. This suggests that girls start off at a disadvantage in school science (particularly in physics, where the sex difference is substantial at ten years old) and never catch up. This disadvantage may take the form of informal learning which boys acquire more easily than girls through their out-of-school socialisation.

If the culture hypothesis holds, the sex differences in science achievement might be expected to differ not only between countries but also between groups in the same country. For example, the norms and expectations in the home and peer group of the child of an unskilled manual worker differ radically (on average) from those of a child of professional parents. And if it is these norms and expectations which produce the sex differences in science achievement, then it is logical to expect some variation in the sex differences corresponding to the variation in family background. Of course, all families within one country are part of the same larger society, and subject to some of the same external pressures and constraints. Nevertheless the dominant influence of the home environment has been demonstrated in many situations (see, for example, Douglas, 1964). A similar influence on sex differences in science achievement is a plausible prediction.

Considerations of time and money made it impracticable to carry out within-country analyses for all fourteen developed countries. Seven countries (Australia, England, Hungary, Italy, Japan, Sweden and the USA) were chosen for detailed study. This choice was made so as to maintain the geographical and cultural spread of the survey, and to eliminate countries with small or unreliable samples. The remainder of this chapter is mainly concerned with results from these countries.

A pupil's family background is traditionally represented by his or her father's occupation. Although clearly not affording a complete description, this measure does give some indication of the different social environments and peer groups within a country. In the IEA studies, father's occupation was recorded on a scale which varied slightly from country to country. But in all cases the lower categories referred to unskilled workers' occupations, while the upper categories referred to professional and managerial

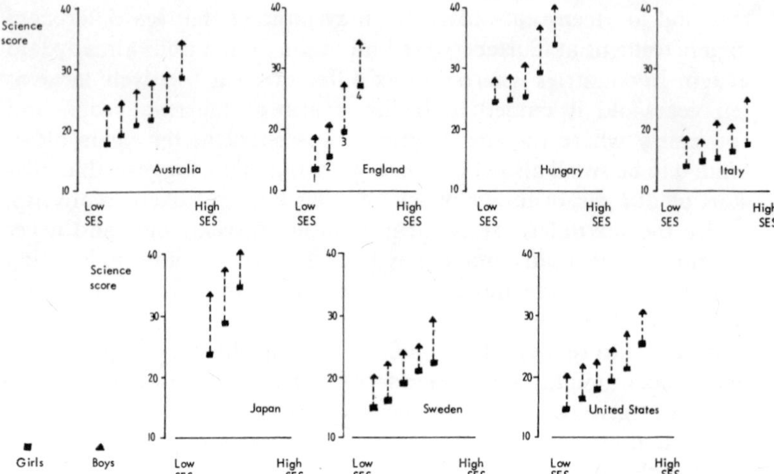

Fig. 2.2 The science achievement of girls and boys from different socio-economic backgrounds. SES; socio-economic status.

occupations. In England the four main categories were (1) unskilled or semi-skilled manual workers, (2) skilled manual workers, (3) white-collar workers, (4) professional and managerial.

Figure 2.2 shows the mean science achievement for girls and boys in each fathers' occupational category. Categories with fewer than 100 pupils of each sex have been omitted. The occupational categories are not truly comparable across countries, so it is misleading to make international comparisons here. But within each country we can see that, as expected, children from higher socio-economic groups did better in science than children from lower socio-economic groups. However, the variations were generally similar for both sexes. The difference in science score between girls and boys in the same social group was approximately constant. Girls in some occupational categories did better than boys in other categories, but within each category the influence for or against science achievement seemed to be independent of sex. In particular, with the possible exception of Japan, there was no evidence of more egalitarian expectations or a narrowing of the achievement gap in the higher socio-economic groups. In England and Hungary the reverse was true, and if anything the gap was narrower in the lower socio-economic groups. But these variations in the gap between the sexes are fairly small and probably due to random effects. In England both the sex differences and the socio-economic differences were highly significant (beyond the 0·1 per cent level), but the interaction between sex and socio-economic group was not significant (in other words the variation in the sex difference from one socio-economic group to another was probably attributable to the chance way the samples were chosen).[8]

The within-country examination of the culture hypothesis thus largely confirmed the results of the between-country analysis. There were variations between the science achievement of different cultures and sub-cultures, and these variations could be attributed to varying societal expectations and pressures. But the culture hypothesis suggested that the relative expectations for girls and boys would vary between cultural groups and produce varying sex differences in science achievement, and this was not found. Societal influences seemed to affect girls and boys in similar ways, and the sex difference in science achievement was more or less constant in the different social groups.

One possible interpretation of this result is that the culture hypothesis is wrong, and that societal expectations and pressures do not account for girls' under-achievement in science. But it must be remembered that the variation in expectations between the cultural and sub-cultural groups considered here was not measured, only assumed. Another interpretation of the results would be that all the cultures and sub-cultures considered here had fundamentally similar expectations for girls in science and manifested fundamentally similar cultural pressures. This position is argued by Elinor Kelly in Chapter 4. Clearly it cannot be ruled out, and much more rigorous tests are required before the culture hypothesis can be rejected. But the culture hypothesis has certainly not been strengthened by this investigation. It is curious that the proportion of science students who are women should vary so widely between the countries considered (presumably owing to social pressures) while the gap between the sexes in science achievement remains impervious to these influences.

The school hypothesis

The school hypothesis suggests that science is presented in schools in a way more suited to boys than to girls. This implies that girls and boys respond differently to school conditions, and that if science were presented in a different way girls would achieve better. If the two sexes enter school with different past experiences, different knowledge, different interests, different attitudes and different expectations it is by no means obvious that the same treatment will have the same effect on them. It may be that treating girls and boys identically in school will serve to accentuate rather than diminish the existing differences. This result is all the more likely when the treatment builds upon a foundation of knowledge, experience and attitudes towards science which is present for one sex but absent for the other.

Previous work on the IEA surveys has shown disappointingly few significant or consistent school effects (see Comber and Keeves, 1973). Other studies have been equally inconclusive (see Welch, 1972, for a review). However, it might be that different relationships were present for girls and boys, and that these effectively cancelled each other out when the sexes were considered together. The school hypothesis was tested by examining the correlations between science achievement and various school level variables for girls and boys. Correlations which differ markedly either in sign or in magnitude for the two sexes may indicate the operation of different processes.

School organisation and practice differ widely from country to country. But similar remarks contrasting girls' diligent and conscientious learning style with boys' more erratic and unpredictable learning style can be found in the educational literature of many countries. If this pattern exists on an international scale, then it is possible that teaching styles affect the sexes differently on an international scale—and perhaps produce sex differences in achievement on an international scale. Sex-dependent factors such as girls modelling themselves on a female teacher or being more confident in girls-only schools might also be expected to show similar effects cross-nationally.

Several studies in England (DES, 1975; Ormerod, 1975; see also Chapters 7 and 14) have indicated that girls have better attitudes towards science and are more likely to continue the study of science in single-sex schools than in co-educational schools, even allowing for selectivity. However in the IEA study the mean science achievements of fourteen-year-old girls were virtually identical in both types of school when allowance was made for selection procedures. This was true for England and for other countries. This result indicates that it is important to distinguish between achievement in science and attitudes towards science when considering the effects of co-education.

The effects of the sex of the science teachers in a school on girls' and boys' achievement are shown in Table 2.2. The top line shows the correlation between science achievement and the proportion of women science teachers in the school.[9] The figures for England show

Table 2.2. Correlations between the proportion of women science teachers in a school and science achievement.

	Australia	*England*	*Hungary*	*Italy*	*Japan*	*Sweden*	*USA*
Girls							
Zero order correlations	0·27*	0·24*	0·05	0·15*	–0·06	–0·01	0·06*
Partial correlations:							
controlled for WK	0·06	–0·10	–0·07	0·08	†	–0·02	–0·21*
controlled for WK and TOS	0·01	–0·20*	–0·07	0·07	†	–0·03	–0·24*
Boys							
Zero order correlations	–0·14*	–0·20*	0·15*	0·18*	–0·21*	0·20*	–0·16*
Partial correlations:							
controlled for WK	–0·08	–0·15	0·03	0·08	†	0·11	–0·06
controlled for WK and TOS	–0·06	–0·18*	0·04	0·04	†	0·11	–0·07

Notes

WK word knowledge; TOS type of school.

Positive correlations indicate better achievement with more woman science teachers; negative correlations indicate better achievement with more men science teachers.

* Significant beyond the 5% level.

† Japan did not administer the word knowledge test, so it was not possible to compute partial correlations controlled for word knowledge for Japanese pupils.

that, overall, girls do better with more women teachers. However, it is well known that in England in 1970 the more able pupils tended to be in single-sex grammar schools, where they were taught by teachers of their own sex. When this is taken into account by controlling for the average ability (as represented by the word knowledge scores, WK) of the pupils in a school and for the type of school (TOS), the partial correlations between science achievement and the proportion of women science teachers show that both girls and boys in England did somewhat better in science with more men teachers. In most other countries the sex of the teacher has no effect on pupils' achievement after allowing for word knowledge and type of school, but in the United States girls again did better with a higher proportion of men teachers in the school. This result runs counter to the usual suggestion that girls will achieve better with more women science teachers to act as role models. However, it must be noted that there is no information in this study on how the male and female teachers were divided between physical and biological sciences, nor on which pupils were actually taught by which teacher. Moreover, as several contributors to this volume point out, some female science teachers may not provide an attractive role model for their teenage pupils!

Other organisational factors and teacher characteristics were treated in less detail than the sex of the teacher, but again failed to reveal any substantial differences between the sexes. The same was true for teaching methods. This international study revealed no great disparities between the sexes in their responses to science teaching and there was no evidence of girls and boys having different learning styles or different reactions to similar treatment.

In most of the countries boys had, on average, attended more physics and chemistry classes than girls, although girls might have attended more biology classes. When this was taken into account the sex differences in chemistry and physics achievement were slightly reduced, although by no means eliminated. Boys' advantage over girls in biology increased when the amount of instruction was considered. Chemistry emerged as the science subject most affected by instruction in school, but only in physics did instruction appear to be more important for girls than for boys. This may have been because girls' background was most lacking (compared to boys') in physics, and may indicate a need for remedial work with girls in physics.

The school hypothesis received little support from this investigation. There were few indications that different approaches to science benefited one sex rather than the other either within individual countries or consistently across countries. On the contrary, the general picture was that school experiences had similar connotations for both sexes. Correlations between science achievement and school-based factors were generally small and inconsistent from country to country, and where relationships did exist they were generally similar for both sexes. Few of the school effects were large enough or consistent enough to serve as a basis for policy changes. Thus the specific hypothesis that science is presented in schools in a way more suited to boys than girls was not supported by the data.

The attitude hypothesis

It is often suggested that girls perform less well than boys in science because girls have less favourable attitudes towards science. This is

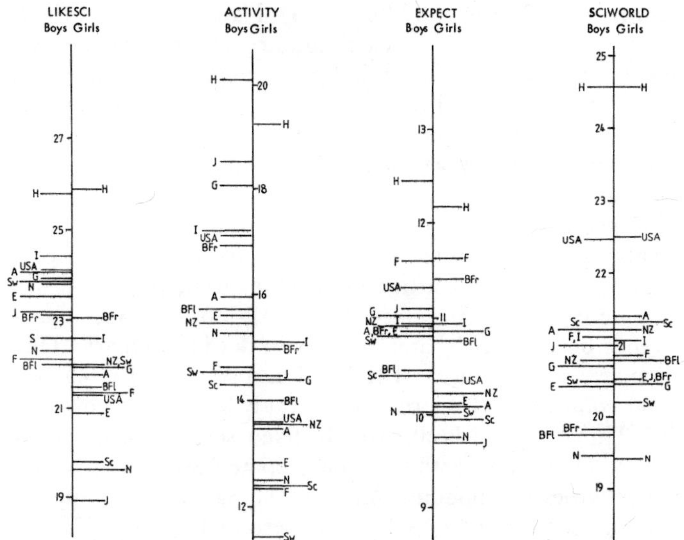

Fig. 2.3 Mean scores on attitude to science scales.

Notes

(a) A – Australia; BFl – Belgium (Flemish); BFr – Belgium (French); N – Netherlands; NZ – New Zealand; Sc – Scotland; Sw – Sweden; USA – United States.

(b) The scale for each attitude is proportional to the average within-country standard deviation of scores. This enables between-attitude comparisons of dispersion to be made by eye.

the attitude hypothesis. A close link between attitudes and achievement is certainly plausible. If pupils work more enthusistically at subjects they enjoy, this extra work probably results in better achievement. If pupils have science-based hobbies, the knowledge acquired in the hobby probably helps in school science. The attitude hypothesis is practically important to the extent that, if girls' attitudes to science could be improved, their achievement might also improve.

In examining the connections between attitudes to science and achievement in science, four attitude scales were derived from the IEA data. These measured linking for science (LIKESCI, e.g. 'I like reading about science'), science-based activities (ACTIVITY, e.g. look at the moon or planets through a telescope), expectations in science (EXPECT, e.g. 'Science is a very difficult subject') and opinions about the effects of science (SCIWORLD, e.g. 'Science helps to make the world a better place to live in'). Details of the scales are given in Kelly (1978, p. 146). Figure 2.3 shows the mean score for girls and boys in each country on each of the attitude scales. Internationally, boys indulged in many more science-based activities than girls, had a substantially greater liking for science, and slightly greater expectations of success in science. There was little overall difference between the sexes in their view of science in the world.

This overall picture of sex differences in attitudes to science conceals quite pronounced variations between countries. The strongly positive mean attitudes of Hungarian pupils on all four scales were outstanding (see Fig. 2.3). So too was the small difference between girls' and boys' attitudes in Hungary. This was particularly striking on the LIKESCI and ACTIVITY scales, since in other countries girls scored considerably lower than boys on these scales. Hungarian girls apparently liked science as much as did Hungarian boys, and both sexes participated extensively in science-based activities and hobbies. Although Hungary was unique among the countries surveyed in achieving these high scores, this result is important in showing that in some circumstances science can appeal to girls. It is, of course, worth noting that Hungary was the only East European country in the IEA science survey, and that East European countries lay a heavy emphasis on science, both in the school curriculum and in its cultural and recreational aspects. Similar efforts to emphasise science in the USA in the post-Sputnik era are perhaps reflected in American pupils' positive attitudes towards science in the world, but not in their scores on the other attitude scales. At the other end of the spectrum, Japanese pupils showed the largest sex differences on all four attitude measures. Although Japanese boys had attitudes similar to boys in other countries, Japanese girls had poor attitudes to science, particularly on the LIKESCI and EXPECT scales.

The size of the sex differences in attitudes to science varied considerably more between countries than did the size of the sex differences in achievement. Arguing in the same way as for the cultural hypothesis, this variation can be attributed to varying cultural expectations and pressures. Sex differences in attitudes were apparently more susceptible to cultural influences than were sex differences in achievement. Moreover girls' attitude scores varied more between countries than boys' attitude scores, which suggests that girls' attitudes to science were particularly malleable, and strongly affected by the prevailing expectations in each country. This malleability is important because, in all the countries studied,

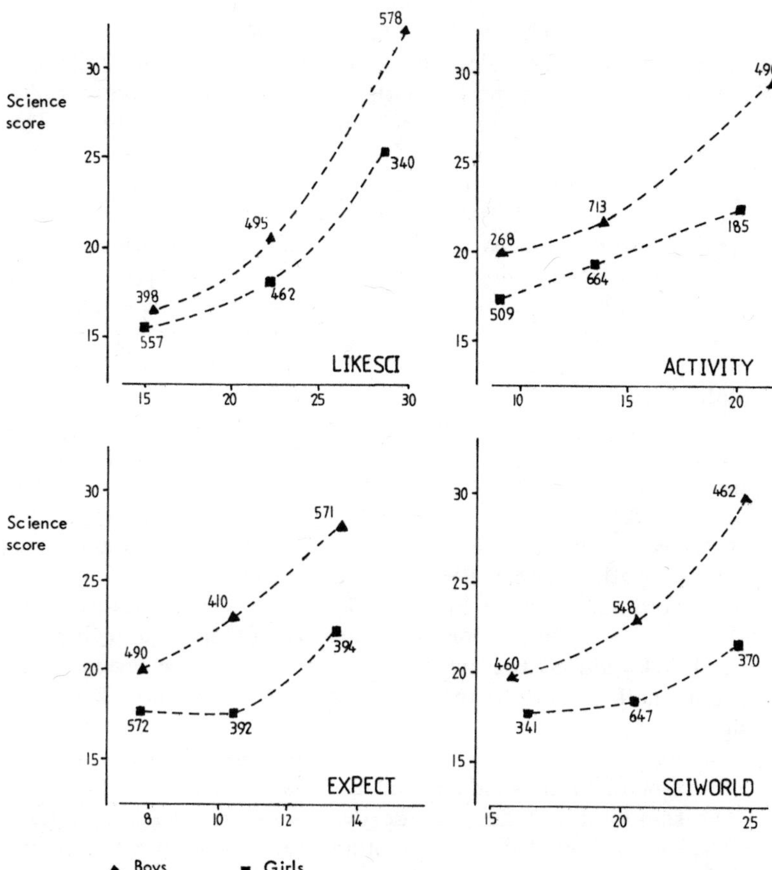

Fig. 2.4 The relationship between attitudes to science and achievement in science in England.
Note Figure by each point gives the number of pupils represented by that point.

pupils with favourable attitudes towards science tended to achieve better in science than pupils with less favourable attitudes. Attitudes towards science were significantly related to achievement in science even when ability (as measured by the word knowledge test) had been taken into account. The general scale, LIKESCI, was more strongly related to achievement than were the other attitude scales, and the correlations were usually somewhat larger for boys than for girls (see Kelly, 1978, p. 97, for details).

The science achievements of English girls and boys with the same score on the attitude scales are compared in Fig. 2.4. This illustrates the points already made. For both sexes, pupils with good attitudes achieved better average scores than pupils with poor attitudes. The effect was strongest for LIKESCI and was generally more pronounced for boys than for girls. Moreover boys with a certain score on the attitude scale consistently outperformed girls with similar attitudes. Indeed, for three of the four attitude scales girls in the top attitude grouping had average science scores approximating to those of boys in the middle attitude grouping. Of course, the sexes were not equally distributed along the curves, with girls being disproportionately at the low end and boys disproportionately at the high end. Nevertheless even those few girls who liked science, had science-based hobbies, expected to do well in science or considered science a force for good in the world did not achieve as well as similarly motivated boys. Indeed, the sex difference in achievement was generally greater among the pupils with the most positive attitudes to science than among the less enthusiastic pupils. The results for other countries were similar. However, in most countries the overall sex difference in science achievement was reduced when allowance was made for attitudes. In England allowing for LIKESCI reduced the average sex difference in achievement by nearly half—although the remaining sex difference was still significant beyond the 0·1 per cent level. Attitudes towards science were not the only, or even the crucial, determinant of science achievement, but they did play an appreciable role.

So some parts of the attitude hypothesis were well supported. Girls *did* have less favourable attitudes towards science than boys, and there *was* a connection between attitudes and achievement: good attitudes were associated with high achievement. Moreover attitudes, and the sex difference in attitudes, varied from country to country, which suggested that girls' attitudes were susceptible to improvement. However, boys' achievement was more highly correlated with attitudes than was girls', and boys achieved better in science than did girls with equally favourable attitudes.

Reassessment

This study began with a model of the development of sex differences in science achievement. Two main processes were postulated. It was suggested that cultural expectations and pressures acted on girls and boys whose abilities and interests were originally similar, and produced sex-typed scientific attitudes and achievements. Then the model suggested that the sex differences in attitudes and achievement were further accentuated by the school, which used similar methods for girls and boys when the two sexes would respond better to different methods.

From this model three hypotheses of the development of sex differences in science achievement were derived, based on culture, school and attitudes. Of these only the attitude hypothesis has received any empirical support from this study—and in the original model attitudes and achievements were both treated as outcomes of the cultural and school processes. The culture hypothesis was not strengthened by this investigation, although the evidence was not such as to allow it to be conclusively rejected. Sex differences in science achievement were certainly present among fourteen-year-old pupils, but they were uniform across cultures and sub-cultures, and there was little sign of specific cultural expectations and pressures influencing their development. Nor was the school hypothesis strengthened. Although the sex differences were larger among fourteen-year-olds than among ten-year-olds, there was little evidence that the school operated differently on girls and boys. The relationships between school factors and science achievement were similar for both sexes. It is possible that these negative results were due to inadequate conceptualisation and measurement of the relevant variables. But intellectual parsimony suggests that alternative viewpoints should be preferred to the further elaborations and qualifications necessary to rescue the original hypotheses.

The remainder of the theoretical section of this book is taken up with such alternative viewpoints. In Chapter 3 Jeffrey Gray suggests that the constant gap between girls and boys in science achievement is the result of a genetically based sex difference in spatial ability. In Chapter 4 Elinor Kelly argues that the apparent cultural differences between the countries studied are minor compared to the underlying similarity in their attitudes towards women and girls, particularly at the time of puberty. My own reinterpretation of the data, in terms of a cognitive developmental theory of the acquisition of sex roles, is presented in Chapter 5. And in Chapter 6 Esther Saraga and Dorothy Griffiths argue that the masculine domination of science is a result of the masculine domination of society, and will only be altered by a radical transformation of patriarchal capitalist society.[10]

Notes

1 This chapter is based on a paper presented at the World Congress of Sociology, Uppsala, Sweden, in August 1978.

2 The IEA (International Association for the Evaluation of Educational Achievement) is an organisation convened under UNESCO auspices to study school achievement cross-nationally. Its first study was in mathematics (Husén, 1967), and this was followed by surveys in science, literature, reading, English and French as foreign languages, and civic education (see Walker, 1976). A second mathematics study is currently under way.

3 I am grateful to the IEA council for allowing me access to their data, to the Spencer Foundation for a grant which enabled me to spend a year in Sweden analysing that data, and to the members of the Institute for International Education, University of Stockholm, for their hospitality and help during that time.

4 The standardised sex difference (SSD) was computed by dividing the difference between boys' and girls' mean scores in each country by the standard deviation of scores in that country. Thus the difference that an educational system produced between boys' and girls' results is expressed as a percentage of the spread (standard deviation) that system produced among all pupils.

5 See the Statistical Appendix for a discussion of standard deviation.

6 In considering these figures it should be remembered that the average difference in score between two individuals chosen at random from a normally distributed population is 113 per cent of the standard deviation. Although the sex difference in science achievement is pronounced, it tells us very little about the likely score of any individual girl or boy.

7 The scores in the four parts of the science test were cumulative, and pupils who scored above average on one part tended to score above average on the other parts as well. This meant that the standard deviation of scores on the total test was considerably larger than the standard deviation of any part of the test. The sex differences were cumulative in a similar way, and the sex difference in the total test (48 per cent of the standard deviation) is larger than the simple average of the sex differences in the component parts of the test.

8 See the Statistical Appendix for a discussion of significance testing.

9 See the Statistical Appendix for a discussion of correlation and partial correlation.

10 I would like to thank Judy Samuel for reading and commenting on an earlier draft of this chapter.

References

Comber, L. C., and Keeves, J. P. (1973), *Science Education in Nineteen Countries*, Almqvist & Wiksell, Stockholm.

Department of Education and Science (1975), *Curricular Differences for Boys and Girls*, Education Survey 21, HMSO, London.

Douglas, J. W. B. (1964), *The Home and the School*, MacGibbon & Kee.

Husén T. (ed.) (1967), *International Study of Achievement in Mathematics*, Almqvist & Wiksell, Stockholm.

Kelly, A. (1978), *Girls and Science: an International Study of Sex Differences in School Science Achievement*, Almqvist & Wiksell, Stockholm.

Ormerod, M. B. (1975), 'Single sex and co-education: an analysis of pupils' science preferences and choices and their attitudes to other aspects of science under these two systems', paper presented at *Girls and Science Education* conference, Chelsea College, London, 1975.

Peaker, G. F. (1975), *An Empirical Study of Education in Twenty-one Countries: a Technical Report*, Almqvist & Wiksell, Stockholm.

Walker, D. A. (1976), *The IEA Six Subject Survey: an Empirical Study of Education in Twenty-one Countries*, Almqvist & Wiksell, Stockholm.

Welch, W. W. (1972), 'Review of research, 1968–69, in secondary level science', *J. of Research in Science Teaching*, Vol. 9, 97.

3

A biological basis for the sex differences in achievement in science?

J. A. GRAY

A striking feature of Alison Kelly's findings reported in Chapter 2 is that the differences between the sexes on the four tests of science achievement were remarkably constant across the fourteen countries for which data are presented. (See Table 2.1.) This constancy occurs in spite of great differences in the average level of achievement between the countries concerned. Thus if the sex difference in science achievement is due to environmental factors, they are present in all the countries studied and do not vary much from one to another. This does not exclude environmental factors, but it is not easy to see what they may be, as Kelly's own discussion makes clear. In contrast, if the sex difference is due to biological factors, constancy in their effect across a range of different cultures is to be expected.

The first place to start looking for biological factors is in the action of genes. The appropriate data to determine whether the sex difference in science achievement is due in any significant degree to genetic effects have not, to my knowledge, been reported. These kinds of data do exist, however, for a particular kind of intellectual ability which is strongly involved in much scientific thinking.

This ability is termed 'visuo-spatial' or simply 'spatial' ability. The kind of task which is used to measure it involves the perception and manipulation of spatial relationships, especially in the visual mode.[1] There is a well documented sex difference in tasks of this nature: males regularly outperform females (Maccoby and Jacklin, 1975; Wilson and Vandenberg, 1978; Yen, 1975). There is evidence that tests of spatial ability can predict success in school geometry, quantitative thinking and performance on mechanical tasks.[2] Werdelin (1958) showed that spatial ability forms one part of the structure (as defined by factor-analytic methods) of mathematical thinking, which is, of course, important throughout science. And physical science specialists have frequently been shown to score higher on tests of spatial ability than arts, social science or biological science specialists (Butcher and Pont, 1969; Child and Smithers, 1971; Hudson, 1966; Lewis, 1964; Roe, 1952).[3] Thus it is possible that the sex difference in science achievement documented in Chapter 2 reflects the sex difference in spatial ability already known to exist.

Intuitively one would expect the requirement for spatial thinking to be greatest for physics and least for biology, with chemistry somewhere in between. This expectation dovetails with Kelly's observation (Table 2.1) of increasing sex differences as one goes from biology through chemsitry to physics. Note also that the difference between the sexes in spatial ability is of the same order of magnitude as Kelly (Table 2.1) reports for the two tests (physics and practical) which produced the biggest sex differences in science achievement, i.e. 0·5–0·7 of a standard deviation (see Yen, 1975, for a representative recent report).

What, then, is the evidence that the sex difference in spatial ability is due to biological factors, and more specifically to genetic factors?

Presumptive evidence that biological factors are indeed at work comes from observations that similar sex differences exist in other species than our own. The only one, however, for which good data are available is the rat. Male and female rats do not generally differ in their performance on most problem-solving tasks, but males are consistently superior in solving complex mazes (Buffery and Gray, 1972). This may be connected with the fact that, in the wild, male rodents have larger home ranges and explore more than females (Brown, 1966), a sexual dimorphism which appears also among primates, human hunter-gatherers and contemporary American children (Harper and Sanders, 1978).

We critically need data on the spatial abilities of non-human primates in order to determine the likelihood that the differences between men and women in this kind of performance are part of a pre-human biological heritage. The only such data known to me are from Van Lawick-Goodall's (1968) monograph on the wild chimpanzee. It appears from her observations that aimed throwing of objects (an activity which involves spatial skills to a high degree) is almost exclusively a male occupation in this species, as in our own (Gesell, 1940; Gesell and Ilg, 1946).

It would also be presumptive evidence for the operation of biological factors if it could be shown that the sex difference in spatial ability commences at a very early age. It is, however, difficult to conduct experiments adequate to establish this point; reviews of the relevant literature have reached divergent conclusions (Buffery and Gray, 1972; Fairweather, 1976).

It should be noted, however, that if it proves to be the case that the difference in spatial ability emerges only in adolescence, this would not constitute evidence *against* a genetic hypothesis. There are many examples of genetically controlled conditions which start only late in life. For example, the greater height of boys than girls does not become established until about fourteen (Tanner, 1960), and the resemblance between the IQs of adopted children and those of the biological parents from whom they are separated grows steadily from infancy to a maximum at age seven (Honzik, 1957). Thus the sex difference in spatial ability, which is clearly established at least by age fourteen (Yen, 1975), could be due to genetic factors even if it is not present at earlier ages.

A third argument which can be used to support the general plausibility of a genetic basis to the sex difference in spatial ability comes from evidence that individual differences in this ability *within* the sexes are heritable to a degree which accounts for about 50 per cent of the variance in test scores (Cattell, 1971, p. 281; Martin and Eaves, 1977).[4] This evidence shows that spatial ability is under genetic control, so that the sex difference in this ability might also be. However, it must be stressed that it is entirely possible for differences in spatial ability within each sex to be largely genetic, while those between the sexes are entirely environmental in origin: the argument is one of plausibility only. If we are to advance beyond plausibility it is necessary to postulate a specific and testable model for the inheritance of the sex difference in spatial ability. One such model has been proposed by O'Connor (1943) and Stafford (1961) who suggested that there is a recessive gene for superior spatial ability carried on the X chromosome.[5] A son receives his single X chromosome from his mother; of her two X chromosomes, a daughter receives one from her mother and one from her father. This results in a particular pattern of expected correlations between parents and offspring for characteristics determined by X-linked genes. It can be shown that, under certain assumptions, there should be no relationship between the spatial ability scores of father and son; a small relationship between the scores of mother and daughter; and a larger relationship between the scores of cross-sex pairs, i.e. mother–son and father–daughter.[6] Similar arguments lead also to an expected pattern of correlations among siblings: the sister–sister correlation should be greater than the brother–brother correlation, and this in turn should exceed the sister–brother correlation (Mather and Jinks, 1971; Yen, 1975).

As reviewed by Buffery and Gray (1972), early studies of parent–offspring correlations on tests of spatial ability tended to support these expectations (see Table 3.1), as did the first study of sibling correlations (Yen, 1975). This constitutes strong evidence for the specific model of inheritance via an X-linked recessive gene. Moreover the pattern of parent–offspring correlations observed in these studies is very difficult to account for on any obvious environmental hypothesis. In particular it more or less rules out any direct modelling by children of their parents' behaviour as an explanation for the sex difference in spatial ability, since this would predict higher same-sex than cross-sex correlations. In addition, the low correlation between fathers and sons (non-significantly different from zero in these early studies) makes it difficult to see how the general home conditions could have had much effect on performance on the tests used.

Further evidence supporting the hypothesis of an X-linked gene was provided by Bock and Kolakowski (1973). These workers looked at the distributions of scores on a test of spatial visualisation for males and females separately. The female distribution was unimodal, that is, had one peak; but the male distribution was bimodal, with an anti-mode (that is, a trough) separating its two peaks. The bimodality of the male distribution may be interpreted as representing two separate distributions, one corresponding to the recessive allele for superior visuo-spatial performance, the other to the allele for inferior performance. The anti-mode in Bock and Kolakowski's (1973) data was near the fiftieth percentile, i.e. about half the male sample scored above it and half below. Thus, on the 'two gene' interpretation, half the males were carrying the 'good' gene for spatial ability. It can be shown that this hypothesis requires the 'good' gene to be present in about 25 per cent of the female sample.[7] Bock and Kolakowski (1973) tested this inference by dichotomising the female distribution at a value corresponding to the fiftieth percentile for the males. As predicted, about a quarter of the females scored above this value. Other workers have similarly reported that about 25 per cent of females score above the median (fiftieth percentile) value for males on spatial ability tests (review by Bock and Kolakowski, 1973); and Yen (1975) and Loehlin *et al.* (1978) have confirmed Bock and Kolakowski's (1973) essential findings with respect to the distribution of male and female scores on tests of a spatial factor. The results of all three experiments are consistent with the suggestion that the recessive allele determining superior spatial ability occurs on about half of all X chromosomes.

There are, however, two major problems with the X-linked model. First, the data from individuals with Turner's syndrome, i.e. phenotypic females[8] with a single X chromosome, are at first sight at variance with the model. Since these individuals have a single X chromosome, the X-linked model predicts that their performance on spatial tests should resemble that of males (Garron, 1970). In fact, however, women with Turner's syndrome not only fail to demonstrate superior performance on spatial tests but are specifically impaired on them (Money and Erhardt, 1968; Nyborg, 1976). These findings add to the general likelihood that genetic factors, and in particular those carried on the sex chromosomes, influence the sex difference in spatial ability; but they are incompatible with the specific X-linked hypothesis that we have considered so far.

There is, however, a likely explanation for the findings in Turner's syndrome which can be fitted quite well with the X-linked hypothesis. It is known that in animals several sexual dimorphisms depend critically on the action of testosterone, a male hormone or 'androgen' released from the testes, during early development (see Gray and Drewett, 1977, for review). This hormone appears to act in the pre-natal and early post-natal periods to alter the organisation of neural systems so that they respond differently to relevant inputs in adulthood. Effects of this kind have been described for the control by the brain of gonadal hormones, and for reproductive and aggressive behaviour. It should be stressed, however, that the relevant experiments have largely been carried out in laboratory rodents, and it is uncertain to what extent similar mechanisms operate in primates and humans (Karsch *et al.*, 1973; Gray and Drewett, 1977). Nonetheless it is of interest that the sex difference in spatial ability in the rat (Buffery and Gray, 1972) is dependent on the presence of testosterone in the neo-natal period (Stewart *et al.*, 1975). This finding raises the possibility that the human sex difference in spatial ability is similarly dependent on early hormonal influence.

If so, the postulated recessive gene which determines superior spatial ability may require for its proper activation a sufficient amount of circulating androgens during early life. In normal males this would be provided, of course, principally by testosterone. There are, however, other sources of androgen. The female ovary provides some, and so do the adrenal glands in both sexes. Individuals with Turner's syndrome, however, have neither testes nor properly developed ovaries (Federman, 1967). If we suppose, therefore, that adrenal androgens are insufficient on their own to activate the hypothetical androgen-sensitive gene, we would have an explanation of the impairment in visuo-spatial ability observed (alongside unimpaired verbal IQ) in patients with Turner's syndrome.

It would be possible to test this hypothesis directly in the rat if it could first be shown that the sex difference in spatial ability in this species is also dependent on the operation of an X-linked gene. To my knowledge, experiments directed towards the latter point have not been performed. There are, however, data from another human pathological condition which support a role for testosterone in the development of visuo-spatial ability. This is the testicular feminisation syndrome (Federman, 1967; Ohno, 1971). The individuals concerned have an XY (normal male) genotype, and normal testes. But there is no response to testosterone in the tissues of the body, and the individual develops into a phenotypic female. Thus on the hypothesis that, for the full development of visuo-spatial ability, it is necessary for the developing cells of the brain to undergo an organisational influence from androgens during early life, it can be predicted that there will be a specific impairment in spatial ability in the testicular feminisation syndrome. This is indeed so (Masica *et al.*, 1969).

The data discussed so far, then, can be accommodated within the framework of a hypothesis which attributes the sex difference in spatial ability to an X-linked recessive gene for superior performance which requires neo-natal or pre-natal androgens for its full expression. But there is a second problem with the X-linked hypothesis. Recent experiments have failed to confirm the early reports of parent–offspring correlations fitting this hypothesis. These experiments are summarised in Table 3.1, which is a more complete version of a table given by DeFries, Vandenberg and McClearn (1976). They have been well conducted, and have used larger samples than earlier studies. They disconfirm the predictions of the X-linked model at virtually every point. In addition, McGee (1978) and Loehlin *et al.* (1978) have both failed to confirm Yen's (1975) findings with sibling correlations.

Faced with such contradictions in the data, there are obvious possible responses. One can conclude that the game is not worth the candle; one can cite the results one likes and pick holes in the others; or one can seek for reasons why the results of the different experiments diverge. If it can be done, the latter course of action is surely to be preferred.

Table 3.1 Parent–child correlations on tests of spatial ability (summary table)

		Correlations			
Study	*Test*	*Father–son*	*Mother–daughter*	*Mother–son*	*Father–daughter*
Bock	Embedded Figures	–0·05	0·36	0·18	0·49
Stafford	Identical Blocks	0·02	0·14	0·31	0·31
McGee	Hidden Patterns	0·07	0·35	0·23	0·39
Park	Card Rotations	0·12	0·61	0·36	0·54
Bock *et al.*	Spatial Visualisation	0·15	0·12	0·20	0·25
DeFries E	Mental Rotations	0·15	0·32	0·16	0·23
Loehlin	Cube Comparisons	0·16	0·19	0·04	0·17
Corah	Embedded Figures	0·18	0·02	0·31	0·28
Hartlage	DAT Space Relations	0·18	0·25	0·39	0·34
Park	Mental Rotations	0·22	0·46	0·26	0·41
McGee	Mental Rotations	0·23	0·16	0·20	0·17
DeFries J	Paper Formboard	0·24	0·24	0·26	0·21
Spuhler	Card Rotations	0·25	0·03	0·16	0·15
Spuhler	Mental Rotations	0·25	0·04	0·10	0·32
DeFries E	Card Rotations	0·26	0·36	0·19	0·22
DeFries J	Card Rotations	0·26	0·09	0·26	0·11
DeFries E	Spatial Factor	0·27	0·41	0·28	0·32
DeFries E	Paper Formboard	0·27	0·36	0·30	0·40
Loehlin	Card Rotations	0·27	0·40	0·27	0·32
Loehlin	Paper Folding	0·27	0·21	0·24	0·30
Loehlin	Spatial Factor	0·28	0·28	0·34	0·30
DeFries J	Mental Rotations	0·30	0·10	0·14	0·34
DeFries J	Spatial Factor	0·33	0·13	0·12	0·31
Spuhler	Spatial Factor	0·35	0·24	0·13	0·40
Spuhler	Paper Formboard	0·37	0·24	0·12	0·35
Loehlin	Hidden Patterns	0·40	0·22	0·44	0·38
Park	Paper Formboard	0·59	0·57	0·63	0·53

Notes

Studies and numbers (N) of pairs entering the correlations:

Bock (1970), N = 22–26
Bock and Kolakowski (1973), N = 84–115
Corah (1965), N = 30
DeFries, Ashton *et al.* (1976), E = Americans of European ancestry, N = 434–438; J = Americans of Japanese ancestry, N = 130–138.
Hartlage (1970), N = 25
Loehlin *et al.* (1978), N = 192
McGee (1978), N = 99–117
Park *et al.* (1978), N = 99–117
Spuhler (1976), N = 50–64
Stafford (1961), N = 50–64

The rows are in ascending order of the observed father–son correlations.

At this point it is worth going back to a technical proviso. The predicted pattern of parent–offspring correlations depends, *inter alia*, on the assumption that there is no other reason, except the operation of the X-linked alleles, why the spatial test scores of parents and children should resemble each other. To the extent that other factors do cause resemblances to arise the specific predictions made from the X-linked hypothesis may be masked. It is implausible that such factors do not exist; indeed, it can be asserted with confidence that they do.

In a study of the heritability of Thurstone's primary mental abilities, Martin and Eaves (1977) have been able to estimate the proportions of variance in scores on the Spatial Ability factor which are due to (1) genetic effects specific to this ability, and (2) genetic effects common to this and other mental abilities. The latter reflect the dependence of performance and visuo-spatial tests on general intellectual ability (*g*). Similarly, Martin and Eaves are able to separate variance on the spatial factor into two kinds of environmental effects, those which vary between individuals in the same family and those which vary between different families; and then further to subdivide these into components specific to spatial ability and components which operate through *g*.

Now, the X-linked recessive hypothesis applies only to spatial ability (not to *g*), and of course only to that portion of the variance in spatial ability which is genetically controlled. From Martin and Eaves's data, one can estimate this portion to be about 30 per cent of the total variance in performance on tests of spatial ability (their Table 6). It is to this 30 per cent of the variance—and even then, provided that the postulated X-linked recessive gene is the only one which affects specific spatial ability—that the predictions of the X-linked hypothesis apply.

There are two distinctive features of the predictions from the X-linked hypothesis. The first is that the father–son correlation is zero. But both the effect of difference between families in environment (estimated by Martin and Eaves, 1977, as influencing 36 per cent of the variance on the spatial factor) and genetic effects working through general intellectual ability (estimated as influencing 16 per cent of the variance) will tend to produce a non-zero correlation between the scores of fathers and sons on spatial tasks.

The second distinctive feature of the predictions from the X-linked hypothesis is that the cross-sex parent–offspring correlations will be higher than the same-sex correlations. But if there are influences in either the environment or in the genetic component of general intellectual ability which cause fathers and sons (or mothers and daughters) to be more alike in spatial ability than are the cross-sex combinations, these will mask the effects of the postulated X-linked gene. It cannot be asserted that such influences exist; but it would not be surprising if they do, particularly in the case of the environmentally controlled variance.

These arguments show that the really surprising feature of the experiments listed in Table 3.1 is that *any* of them managed to obtain the pattern of parent–offspring correlations (and particularly the zero correlation between fathers and sons) predicted by the X-linked hypothesis. But one cannot defend a hypothesis simply by pointing to all the reasons why its predictions may fail to be confirmed: positive evidence is a *sine qua non* of the experimental method. Nor can one rest content with a pot-pourri of results, some supporting the hypothesis and some not: it is essential that experiments be replicable. The question therefore arises, what happened between the earlier experiments supporting the X-linked hypothesis and the later ones failing to do so?

A glance at Table 3.1 shows that, in fact, the later experiments, while not supporting the X-linked hypothesis, are not failures to replicate the earlier ones. For none of them used the same tests of spatial ability as those used by Stafford, Corah, Hartlage, Bock, or Bock and Kolakowski. Thus the data summarised in Table 3.1 leave open two possibilities. The first is that the earlier results were spurious, chance coincidences between data and theory from experiments using small samples. The second is that some tests produce results which fit the predictions of the X-linked hypothesis, but others do not. If the second account is correct, the arguments about masking influences rehearsed above lead one to specific predictions: tests on which the parent–offspring correlations fit the predictions of the X-linked hypothesis should share strongly in the genetic variance specific to spatial ability, and not much in the between-family environmental variance or in the genetic variance due to *g*; and conversely for the tests on which the parent–offspring correlations do not fit the X-linked hypothesis. I do not know of any published reports which would allow one to test these predictions, and it would be a formidable task to gather the necessary data.

At the moment, then, the status of the X-linked hypothesis is uncertain. There are some reports of parent–offspring correlation, and of the distributions of scores on spatial tasks in the two sexes, which fit it very well and which do not easily lend themselves to other explanations, either genetic or environmental. But there is a formidable array of negative evidence from the recent large-scale studies listed in Table 3.1.

It should be remembered, however, that, if the experimental evidence finally kills off the X-linked hypothesis, this will not by itself do anything to establish an *environmental* account of the sex difference in spatial ability. The X-linked hypothesis is one specific genetic model; but others could be proposed. The probability that a hypothesis will be disproved is in direct proportion to the specificity of its predictions. Genetic models are much more specific in their predictions than are current environmental ones, and therefore more susceptible to sudden death. The real challenge to those who believe that the sex difference in spatial ability is predominently environmental in origin is to come up with a model which is as specific in its predictions as, for example, the X-linked hypothesis which has occupied much of our attention in this chapter.

Whether or not the specific X-linked hypothesis is correct, there are a number of reasons, mentioned above, to suppose that some kind of biologically based account of the sex difference in spatial ability will eventually turn out to be correct. There are two important questions which such a biologically based account must address, and which I shall briefly touch upon in the closing pages of this chapter.

The first concerns the neural mechanisms by which the sex difference is produced. In fact this question must arise whether the sex difference in spatial ability is predominantly environmental or predominantly genetic in origin. For whatever it is that leads the sexes to behave differently, they do so because their brains operate differently (Gray, 1968). From the genetic point of view, genes do not specify behaviour directly. Through a long development chain they specify the organisation and action of systems in the brain, and it is these that control behaviour. Most recent speculations about differences in the brain which may underlie the sex difference in spatial ability have centred on the notion of lateralisation of function, i.e. the notion (well supported empirically) that the left and right sides of the brain are specialised in some degree to perform different functions, and, in particular, that speech is normally subserved principally by the left hemisphere and spatial functions principally by the right (Dimond and Beaumont, 1974; Buffery and Gray, 1972). Several different proposals as to sex differences in the degree or kind of lateralisation of function have been made (Levy, 1972; Sherman, 1967; Buffery, 1970; Buffery and Gray, 1972), attacked (Marshall, 1973; Fairweather, 1976; Sherman, 1977) and defended (Buffery, 1976; Harris, 1978; Goldman, 1976; Sherman, 1977). This is not the place to enter into details concerning this controversial area, in which much experimental work is currently going on.

The second question concerns the survival value, in the Darwinian sense, of the sex difference in spatial ability. For the hypothesis that this sex difference has a genetic basis implies that it possesses a survival value. Here too it will be possible in the present compass only to indicate the nature of arguments that have been developed more fully elsewhere (Wynne-Edwards, 1962; Gray and Buffery, 1971; Harper and Sanders, 1978). Essentially, it has been suggested that the capacity to analyse and utilise spatial information has evolved, at least in some degree, because of its usefulness to a class of behaviour patterns which males typically perform more than females. Among rodents, this behaviour is best described as territorial; among primates, it is concerned with the defence of the group or with hunting prey. Both kinds of behaviour involve aggressive encounters and the exploration of spatially complex environments; the second kind may also involve the manipulation of objects, e.g. the aimed throwing of stones or, in human hunting groups, weapons.

If one takes the question still deeper and asks, why should males be more likely to engage in these activities than females, the answer, following Wynne-Edwards's (1962) important general theory of social behaviour, is that males are more expendable than females. It is the number of females that is the most potent limiting factor on the breeding potential of a group; and breeding potential is, of course, what Darwinian fitness is all about. Thus any genetic mechanism which pushes the males of the group, rather than the females, into risky activities (straying far from home, fighting with others of the same species, defending the group from a predator, attacking a dangerous prey) will have a higher survival value than one which is equally careless of the two sexes.

What, then, should be done about the sex differences in science achievement that Alison Kelly reports? The answer is, nothing. In a recent report on sex differences on fifteen different tests of different kinds of ability, sex emerged to a significant degree in fourteen of them (Wilson and Vandenberg, 1978). On six tests the females had the advantage; on eight the males. Why should we attempt to iron out these differences?

I can think of only one possible reason: that is, if being poor at spatial tasks (and therefore, *ex hypothesi*, at science) leads to poorer social or economic prospects for women. For this to be the case, it would be necessary that society systematically rewarded the skills that males have, rather than those that females have. But this is demonstrably not so. Females, as known for many years (see reviews by Buffery and Gray, 1972; Maccoby and Jacklin, 1975) and as confirmed in the recent Wilson and Vandenberg (1978) study, normally do better than males on tests of verbal ability. It should follow, therefore, that, if society rewards 'male' rather than 'female' skills, the relationship between spatial ability and social class should be higher than that between verbal ability and social class. But exactly the converse is the case: social class is more closely related to verbal than to spatial ability (Bock and Kolakowski, 1973; Jencks *et al.*, 1972, p. 78).

This evidence from the psychometric laboratory is supported by everyday observation. In the United Kingdom science administrators, who mainly exercise the verbal arts, are more highly paid than the spatially skilled scientists whom they administer. The lack of scientists in the higher reaches of industry has often been lamented and blamed for our industrial ills. Thus, if women as a group suffer from social or economic discrimination, this has little or nothing to do with their achievement in science or their relative lack of spatial ability. And when an individual woman (Dorothy Hodgkin is a good example) has outstanding spatial talents, these are duly recognised by society. It is with a clear conscience, therefore, that we may welcome the different patterns of intellectual abilities shown by the two sexes as an addition to diversity in a society which badly needs it, and as an increase in the sum total of talents available.

Notes

1 Examples of tests of visuo-spatial ability would be ones in which the subject is required to say which of a series of shapes is a rotation of a sample shape; to detect a smaller and simpler shape within a larger and more complex design; to adjust a rod to gravitational vertical when a frame around the rod is tilted from vertical; or to solve analogies, presented visually, of the form 'shape A is to shape B as shape C is to which other shape?'

2 The product-moment correlations between scores on spatial ability tests and geometry, quantitative thinking and mechanical tasks are 0·57, 0·69, and 0·47 respectively. Data from the manual for the Differential Aptitude Test (Bennett *et al.*, 1966, cited by Bock and Kolokowski, 1973). See also MacFarlane Smith (1964).

3 This pattern is confirmed in Chapter 11 of the present volume [*Editor*].

4 See the Statistical Appendix for a discussion of variance and the partition of variance.

5 The sex chromosomes are known as X and Y. In mammals, males have one X and one Y chromosome; females have two Xs. A recessive gene is one whose expression is masked if it is paired with a dominant gene on the other chromosome. The Y chromosome is very short, and is not thought to carry any genetic information except that determining the formation of a testis rather than an ovary. The X chromosome is of normal length and carries genetic information specifying many characteristics. Thus a recessive gene on an X chromosome will be expressed in a female only if the other X chromosome bears the same recessive gene, but will always be expressed in a male.

6 Specifically, we have to assume a random mating population, two alternative genes ('alleles') which may occupy a particular place (or 'locus') on the X chromosome, of which one is fully dominant over the other, *and no shared variance other than that due to the operation of these alleles* (a proviso which will become important later; see p. 49). Under these conditions the prediction is that, if q is the frequency in the population of the recessive allele (where q is between 0 and 1), the mother–son correlation = the father–daughter correlation $= \sqrt{q/(q+1)}$; the mother–daughter correlation $= q/(q+1)$, i.e. the square of (and therefore smaller than) the former two correlations; and the father–son correlation = 0.

7 *Ex hypothesi* males carry the recessive gene with a frequency in the population q, the females carry the double recessive on both X chromosomes with frequency q^2 (note that, since q is less than 1, q^2 is less than q). Thus, if half the males have the recessive gene (deduced from the location of the anti-mode at the fiftieth percentile), $q = 0{\cdot}5$ and $q^2 = 0{\cdot}25$.

8 Genotype and phenotype refer to the genetic blueprint and the social end product respectively. Thus a phenotypic female is any person who appears to be and is referred to as a female, whether their genetic make up is the normal XX, the Turner's syndrome XO or in exceptional cases the male XY.

References

Bennett, G. K., Seashore, H. G., and Wesman, A. G. (1966), *Manual for the Differential Aptitude Test, Forms L and M, 4th ed.*, New York: Psychological Corp.

Bock, R. D. (1970), 'A study of familial effects in certain cognitive and perceptual variables', Final Rep. I11. Psychiatr. Train. Res. Grant No. 17–317, Univ. Chicago.

Bock, R. D., and Kolakowski, D. (1973), 'Further evidence of sex-linked major-gene influence on human spatial visualizing ability', *Am. J. Hum. Genet.*, Vol 25, pp. 1–14.

Brown, L. E. (1966), 'Home range and movement of small mammals', Vol. 1, p. 3 in P. A. Jewel and C. Loizos (eds.), *Play, Exploration and Territory in Mammals*, Symp. Zoo. Soc. Lond., Vol. 18, pp. 111–142, London: Academic Press.

Buffery, A. W. H. (1970), 'Sex differences in the development of hand preference, cerebral dominance for speech and cognitive skill', *Bulletin of the British Psychological Society*, Vol. 23, p. 233.

——(1976). 'Sex differences in the neuro-psychological development of verbal and spatial skills'. In R. Knights and D. J. Bakker (eds.) *The Neuropsychology of Learning Disorder: Theoretical Approaches*. University Park Press.

Buffery, A. W. H., and Gray, J. A. (1972), 'Sex differences in the development of spatial and linguistic skills', in C. Ounsted and D. C. Taylor (eds.), *Gender Differences: their Ontogeny and Significance*. Churchill Livingstone, Edinburgh.

Butcher H. J., and Pont H. B. (1969), 'Predicting arts and science specialisation in a group of Scottish secondary school children: some preliminary results' *Scottish Educational Studies* Vol. 1, p. 3.

Cattell, R. B. (1971), *Abilities: their Structure, Growth, and Action*, Boston: Houghton Mifflin.

Child D. and Smithers A. (1971) 'Some cognitive and affective factors in subject choice' *Research in Education*, Vol 5, p. 1.

Corah, N. L. (1965), 'Differentiation in children and their parents', *J. Pers.*, Vol. 33, pp. 300–8.

DeFries, J. C., Vandenberg, S. G., and McClearn, G. E. (1976), 'Genetics of specific cognitive abilities', *Ann. Rev. Genet.*, Vol. 10, pp. 179–207.

DeFries, J. C., Ashton, G. C., Johnson, R. C., Kuse, A. R., McClearn, G. E., Mi, M. P., Rashad, M. N., Vandenberg, S. G., and Wilson, J. R. (1976), 'Parent–offspring resemblance for specific cognitive abilities in two ethnic groups', *Nature*, Vol. 261, pp. 131–3.

Dimond, S. J., and Beaumont, J. G. (1974), *Hemisphere Function in the Human Brain*, London: Elek Science.

Fairweather, H. (1976), 'Sex differences in cognition', *Cognition*, Vol. 4, pp. 231–80.

Federman, D. D. (1967), *Abnormal Sexual Development*. Philadelphia: Saunders.

Garron, D. C. (1970), 'Sex-linked recessive inheritance of spatial and numerical abilities, and Turner's syndrome', *Psychol. Rev.*, Vol. 77, pp. 147–52.

Gesell, A. (1940), *The First Five Years of Life*, London: Methuen.

Gesell, A., and I1g, F. L. (1946), *The Child from Five to Ten*, London: Hamish Hamilton.

Goldman, P. S. (1976), 'Maturation of the mammalian nervous system and the ontogeny of behaviour', in Lehrman, D. S., Hinde, R. A., and Shaw, E. (eds.), *Advances of the Study of Behaviour*, Vol. 7, pp. 1–90, New York: Academic Press.

Gray, J. A. (1968), 'The Lister Lecture, 1967. The physiological basis of personality', *Advancement of Science*, Vol. 24, pp. 293–305.

Gray, J. A., and Buffery, A. W. H. (1971), 'Sex differences in emotional and cognitive behaviour in mammals including man: adaptive and neural bases', *Acta Psychologica*, Vol. 35, pp. 89–111.

Gray, J. A., and Drewett, R. F. (1977), 'The genetics and development of sex differences', in Catell, R. B., and Dreger, R. M. (eds.), *Handbook of Modern Personality Theory*, pp. 348–73, London: Wiley.

Harper, L. V., and Sanders, K. M. (1978), 'Sex differences in preschool children's social interactions and use of space: an evolutionary perspective', in McGill, T. E., Dewsbury, D. A., and Sachs, B. D. (eds.), *Sex and Behaviour: Status and Prospectus*, pp. 61–81, New York: Plenum.

Harris, L. J. (1978) 'Sex differences in spatial ability: possible environmental, genetic and neurological factors', in Kinsbourne, M. (ed.), *Asymmetrical Function of the Brain, pp. 405–522,* Cambridge University Press.

Hartlage, L. C., (1970), 'Sex-linked inheritance of spatial ability', *Percept. Mot. Skills*, Vol. 31, p. 610.

Honzik, M. P. (1957) 'Developmental studies of parent-child resemblance in intelligence', *Child Development*, Vol. 28, pp. 215–28.

Hudson, L. (1966) *Contrary Imaginations*, Penguin.

Jencks, C., *et al.* (1972), *Inequality: a Reassessment of the Effects of Family and Schooling in America*, New York: Basic Books.

Karsch, F. J., Dierschke, D. J., and Knobil, E. (1973), 'Sexual differentiation of pituitary function: apparent difference between primates and rodents', *Science*, Vol. 179, pp. 484–6.

Levy, J. (1972), 'Lateral specialization of the human brain: behavioural manifestations and possible evolutionary basis', in J. A. Kiger (ed.), *The Biology of Behaviour*, Corvallis, Oregon: Oregon University Press.

Lewis D. G. (1964) 'The factorial nature of attainment in elementary science', *Br. J. Educational Psychology*, Vol. 34, p. 1.

Loehlin, J. C., Sharan, S., and Jacoby, R. (1978), 'In pursuit of the "spatial gene": a family study', *Behav. Genet.*, Vol. 8, pp. 27–42.

Maccoby, E. E., and Jacklin, C. N. (1975), *The Psychology of Sex Differences*, London: Oxford University Press.

MacFarlane Smith, I. (1964). *Spatial Ability*. London: University of London Press.

Marshall, J. C. (1973), 'Some problems and paradoxes associated with recent accounts of hemispheric specialization', *Neuropsychologia*, Vol. 11, pp. 463–70.

Martin, N. G., and Eaves, L. J. (1977), 'The genetical basis of covariance structure', *Heredity*, Vol. 38, pp. 79–95.

Masica, D. N., Money, J., Ehrhardt, A. A., and Lewis, V. G. (1969), 'I.Q., fetal sex hormones and cognitive patterns: studies in the testicular feminizing syndrome of androgen insensitivity', *Johns Hopkins Medical Journal*, Vol. 124 (1), pp. 34–43.

Mather, K., and Jinks, J. L. (1971), *Biometrical Genetics*, London: Chapman & Hall.

McGee, M. G. (1978), 'Intra-familial correlations and heritability estimates for spatial ability in a Minnesota sample', *Behav. Genet.*, Vol. 8, pp. 77–80.

Money, J., and Ehrhardt, A. A. (1968), 'Prenatal hormonal exposure: possible effects on behaviour in man', in Michael, R. P. (ed.), *Endocrinology and Human Behaviour*, London: Oxford University Press.

Nyborg, H. (1976), 'Sex chromosome abnormalities and cognitive performance', *Psychological Reports, Aarhus*, Vol. 1, Aarhus, Denmark: University of Aarhus.

O'Connor, J. (1943), *Structural Visualization*, Boston, Human Engineering Laboratory.

Ohno, S. (1971), 'Simplicity of mammalian regulatory systems inferred by single gene determination of sex phenotypes', *Nature*, Vol. 234, pp. 134–7.

Park, J., Johnson, R. C., DeFries, J. C., McClearn, G. E., Mi, M. P., Rashad, M. N., Vandenberg, S. G., and Wilson, J. R. (1978), 'Parent–offspring resemblance for specific cognitive abilities in Korea', *Behav. Genet.*, Vol. 8, pp. 43–52.

Roe A. (1952) *The Making of a Scientist*, Dodd Mead, N. Y.

Sherman, J. A. (1967), 'Problem of sex differences in space perception and aspects of intellectual functioning', *Psychol. Rev.*, Vol. 74, pp. 290–99.

——(1977), 'Effects of biological factors on sex related differences in mathematics achievement', in Shoemaker, J. S. (ed.), *Women and Mathematics: Research Perspectives for Change*, National Institute of Education Washington, D. C.

Spuhler, K. P. (1976), 'Family resemblance for cognitive performance: an assessment of genetic and environmental contributions to variation', PhD thesis. Univ. Colorado, Boulder, pp. 188.

Stafford, R. E. (1961), 'Sex differences in spatial visualization as evidence of sex-linked inheritance', *Percept. Motor Skills*, Vol. 13, pp. 428.

Stewart, J., Skvarenina, A., and Pottier, J. (1975), 'Effects of neonatal androgens on open-field behaviour and maze learning in the prepubescent and adult rat', *Physiology and Behaviour*, Vol. 14, pp. 291–5.

Tanner, J. M. (1960), 'Genetics of human growth', in Tanner, J. M. (ed.), *Human Growth*, Symposia of the Society for the Study of Human Biology, Vol. 3, pp. 43–58, Oxford: Pergamon.

——(1962), *Growth at Adolescence*, Oxford: Blackwell.

Van Lawick-Goodall, J. (1968), 'The behaviour of free-living chimpanzees in the Gombe Stream Reserve', *Anim. Behav. Monog.*, Vol. 1, pp. 161–311.

Werdelin, I. (1958), *The Mathematical Ability*, Lund: C. W. K. Gleerup.

Wilson, J. R., and Vandenberg, S. G. (1978), 'Sex differences in cognition: evidence from the Hawaii family study' in McGill, T. E., Dewsbury, D. A., and Sachs, B. D. (eds.), *Sex and Behaviour: Status and Prospectus*, pp. 317–35, New York: Plenum.

Wynne-Edwards, V. C. (1962), *Animal Dispersion in Relation to Social Behaviour*, Edinburgh: Oliver & Boyd.

Yen, W. M. (1975), 'Sex-linked major-gene influences on selected types of spatial performance', *Behav. Genet.*, Vol. 5, pp. 281–98.

4

Socialisation in patriarchal society

ELINOR KELLY

If we are to discuss the issue of girls and science, as it has been developed in the IEA survey, then we need to develop a cross-cultural perspective. I will focus on the continuity between generations and the communication by adults of attitudes and expectations which are internalised by children. I propose to use the perspective of socialisation to explore the international continuities, and to concentrate on the extent to which 'patriarchy' has endured in each of the countries surveyed.

Patriarchy

'Patriarchy' is a concept which has been used a great deal but rarely defined in feminist literature, because, while arguing that all societies are 'patriarchal', feminists have found no example of a society in which it has been the single organising principle. Patriarchy should be understood as a principle which orders relations between the sexes and between generations on specific lines—it divides home and work into masculine and feminine spheres, and into a hierarchy, with men in more powerful and prestigious positions where they exercise authority over women and children. The male-dominated family in fact lies at the root of patriarchy, because, however rich or poor they may be, men can rely on being the heads of households, in which women submit to their authority. In all societies women's work is anchored in domestic labour, focused on the household and children; and in all societies the early socialisation of children is part of women's domestic labour (Bujra, 1978). So women play a complicated part in maintaining 'patriarchy', because while they are its 'victims', confined to a sphere which is usually narrow and lowly valued, they also perpetuate the existing system through their influence on the next generation.

Why do women perpetuate patriarchal values in their children? The answers to this must vary according to the local conditions. To begin with, many women are not 'consciously aware of being oppressed *as women*, even though their objective situation might be highly circumscribed' (Bujra, 1978, p. 15). Then other women gain mutual support and corporate influence on important decisions through being part of a women's world (Saifullah-Khan, 1976; Cohen, 1978). But, more important, patriarchy cannot be understood by talking only about gender differences. *Generation* differences are extremely important, because many women can look forward to menopause as a time when the rules of sex segregation ease (Bujra, 1978, p. 21; Richards, 1966). Also they can begin to delegate their domestic labour to junior females, i.e. their daughters or the brides of their sons (Croll, 1978, p. 50).

Women have a central role in family prestige and honour. As the mothers of maturing children they protect the reputation of the family into which they have married. The ideal of patriarchy is that the control of a girl's sexuality and fertility should lie first in the hands of her parents, and then in the hands of her husband. Individual choice of marriage partner, freedom to engage in non-marital sexual relations, control of fertility are all issues which relate to the continued functioning of a patriarchal order.

So there is a patriarchal logic behind stereotyped notions of sex roles, and feminists argue that the Western capitalist world still manifests this logic in many ways. The extent to which patriarchy prevails over other social principles obviously varies from one society to another, between the strata of any one society and from family to family. In the most consistently patriarchal situations—for example where Islamic 'purdah' prevails—elders have the monopoly of authority and by puberty the sexes are rigidly separated. In the IEA countries few women are as strictly secluded as in 'purdah', and virtually all children attend school. But it is possible that, while the rules of sex-appropriate behaviour have become less strict and obvious, the principles of patriarchy have spread and adapted into areas of life which could have given women new opportunities. We should remember that in London something like 40 per cent of women may never go out alone after dark (Fremlin, 1979); that in the labour market many women are confined to part-time work and to the 'sex-ghettoes' (Barron and Norris, 1976); and that very few women have been appointed to leading positions in science and technology. There are no formal rules to say women must stay confined, but somehow they are.

The IEA countries all have some degree of commitment to notions of 'equality' of opportunity, and some have enacted legislation making sex discrimination illegal. But changes in law do not change attitudes overnight. In this context we must consider not only the 'formally institutionalised distinctions' which schools make in subject teaching, areas of study and learning activities, but also 'the "hidden curriculum" which transmits to young people a collection of messages about the status and character of individuals and social groups. It works through school organisation, through attitudes and through omissions—what we do not teach, highlight or illuminate, is often more influential as a factor for bias, than what we do' (Byrne, 1978, p. 110).

This distinction is important because it demonstrates that the 'sex ghettoes' which emerge in school subjects will only disappear when the adults concerned realise and alter the messages they are transmitting, often unconsciously, to children. This argument demands not just a formal commitment to equality of opportunity, but an examination of the part which adults play in the socialisation of children, at home and in the school. In our terms, it also demands consideration of the possibility that teachers and parents are still following patriarchal logic.

The roles of adults in socialisation

Anyone who has spent time with small children is aware of how significant sex stereotypes are to them — soon after they can speak they express forcibly the notions that 'I am a girl/boy', or 'only girls do that'. In fact parents are often bewildered by the behaviour of their four-year-olds, who seem rigid and exaggerated in their notions of sex-appropriate behaviour (Maccoby and Jacklin, 1975). These everyday impressions have been confirmed by systematic research. Hutt (1972) found that there were marked differences in the behaviour of two- to five-year-old boys and girls, Goldberg and Lewis (1969) traced the differences back to the play of year-old infants, and Kohlberg (1966) found that in the third year of life children were aware of their own sex labels.

This evidence of sex differences in very small children at first appears damaging for the socialisation theories which argue that sex typed behaviour is learned and not innate. Yet the same studies which have demonstrated sex differences in young children's behaviour frequently make the point that, however early the differences appear, we should not assume that they are innate. From the time of birth girls and boys are treated differently. There is evidence, for example, that boy infants are handled and physically stimulated more than girls, whereas girls are talked to more (Belotti, 1975; Lewis, 1972). Yet, adults are usually unaware of the means by which they are transmitting messages — the tiny gestures, unconscious patternings, tone of voice and body postures, etc., which are as important as the verbal and visual images portrayed in word, objects and pictures (Mischel, 1966).

This argument is frustrating for adults who wish to eliminate sex-typing because it requires self-questioning and observation of seemingly insignificant everyday details. Yet, if we think about Torrance's experiment (1963) in which he was able to record significant improvement in girls' performance with science toys when parents became encouraging; if we pause before we use words such as 'sissy' and 'tomboy' with children, we may see useful lines along which to experiment. To really test out the extent to which sex differences are innate, it is essential to identify and control the messages which children receive and absorb.

Some of the messages are more accessible to research then others—for example, the unconscious patterns which adults reveal in their choice of toys and colour. How many parents will dress their boys in pink, give girls mobiles with yachts, ships or cars on them, dress boys in nurses' kits or encourage girls to play with nuts, bolts or screwdrivers? Adult customers may not be aware of their bias, but sales staff and package designers are very conscious of the principles behind the design and display of toys—they calculatingly conform to and perpetuate traditional stereotypes. In a USA survey 75 per cent of chemistry sets potrayed boys alone on the box, none showed girls alone. Toys such as Meccano which taught engineering principles were marketed exclusively for boys; masculine toys were varied, complex and expensive, feminine toys were simpler, and anticipated passive and solitary activity. The commercial principles were classically patriarchal—toys should be selected on the basis of gender, and greater expense and more attention should be expended on the boys' toys (Stacey *et al.*, 1974).

Surveys of children's books show the same tendencies—overwhelmingly they portray boys as the active characters who go camping, build tree houses, have adventures and are naughty. Girls stay at home, play with dolls and kittens, and help mother. After an international survey Belotti concluded rather despairingly, 'however diligently one searches, it is impossible to find a female character who is intelligent, courageous, active and loyal' (1975, pp. 89–103). In Britain many of the best-selling series, such as *Janet and John* or the Ladybird books have been analysed and heavily criticised for their sexism and perpetuation of sex roles that are so rigidly traditional that they have become caricatures (Children's Rights Workshop, 1976).

Books and toys are only part of the shaping of the social world of a child. Because no viable and continuing alternatives have been developed in most families, early socialisation of children still lies predominantly with mothers—mothers whose own school and family experience is frequently of little help in developing non-stereotypical behaviour in their children. From the research already done it seems that the roots of children's analytic development do actually lie in early childhood, with the attitudes of the parents, particularly the mother; as Maccoby (1963) reported, the mothers of girls who were best at maths and spatial tasks often left their daughters to solve problems by themselves, and those boys who were poorest at maths had been 'overprotected'.

The femininity of primary schools

If we typify the mother-centred home, and the classic continuity between mother and children, as predominantly 'feminine', then we see that the primary school can either perpetuate the femininity of the home, or initiate children into cross-sex activities and attitudes. At present schools seem content to accept existing conventions. Teachers in general are encouraged in their training and by the school environment to believe the argument that 'extreme differences between males and females exist as early as age three. They also appear to believe that it is appropriate to behave differently to the sexes in accord with their "natural" characteristics' (Lobban, 1978, p. 57; see also Blackstone, 1976; Douglas, 1967; Sharpe, 1976).

Teachers usually adopt either these extremely traditional attitudes, or a form of pessimism. They may want things to be different and be discouraged by children's failure to adopt new opportunities while still being fearful of a wave of public disapproval and institutional distaste if they attempt to introduce children to alternative ways of behaving. But change will not come about automatically as a gradual response to social developments. It is exhausting work, requiring conscious effort from the teacher. Lee and Gropper (1974) and Shave (1978) provide some practical guidelines which emphasise the importance of sex role reversals in classroom activities, the need for teachers to be retrained in non-verbal communication, and the avoidance at all times of grouping children by gender.

Continuity with the mother-centred home is perpetuated in many ways in primary schools. The images of textbooks consistently portray women in the archetypal female roles of housewife or mother, only occasionally as teacher, secretary, nurse or shop assistant, and never as scientist or technician (Children's Rights Workshop, 1976). Even more significantly, the primary school environment is overwhelmingly feminine. Primary schools are usually staffed by women and the teachers frequently demand 'obedience, silence, passivity and conformity from their pupils—all features of traditional female behaviour' (Sharpe, 1976, p. 145). There is evidence that boys dominate in taking and being allowed first choice in toys, and take more space for their activities, with the tacit acceptance of the girls and the teachers. Yet teachers also reward the girls for greater tidiness, quietness and obedience, and dislike the boys' mess and noise (Blackstone, 1976).

It is possible that the greater rewards of femininity in primary school actually have adverse effects on some boys, who feel excluded and alienated, and positive effects for girls, who, cocooned in a consistently feminine world, perform better at this stage of schooling than boys (Douglas, 1967; Lee and Gropper, 1974). However, these differences gain a new meaning in secondary school. Many boys, as a result of this process which risks alienating them from school, are actually being trained to a degree of independence and initiative which encourages analytic thinking in later life. By contrast, the girls, by conforming to femininity, are discouraged from originality and experimentation (Maccoby, 1963).

This sex differentiation within primary schools is carried over into science. The Leeds Literature Collective reviewed the representation of girls in primary science books, and found that most of the pictures and most of the pronouns referred to males. One book they reviewed had several illustrations of children performing experiments, but only one of these was a girl; she was shown blowing bubbles—a charming, feminine, but not very scientific, activity. Secondary school science books are similarly biased towards males (Kelly, 1976a; see also Chapter 19). The Leeds Literature Collective concluded that 'most girls will already know (by the time they enter secondary school) that science is unfeminine and that their true vocation lies in the home, with unskilled work if they must work, or in the Arts subjects if they are academically successful' (1973, p. 4). They emphasised that primary school practices confirmed, rather than altered this view: not only are children divided into single-sex groups for many activities, but physical science is often taught by visiting male teachers, and the examples used in early experiments are rarely, if ever, drawn from the female spheres of activities.

Considering the extent of the sex typing to which they are exposed, it is not surprising to find that by that time children leave primary school the differences between girls and boys are marked. Kelly (1976b) listed eight major differences which are relevant to science. At this age girls tend to be more verbal, less independent, more easily discouraged, more conscientious, more interested in people, less interested in science, less experienced in science-related activities and more restricted in their perception of possible future roles than boys.

Secondary school and puberty

When pupils move to secondary school they encounter great changes. Secondary schools are usually bigger and more complex than primary schools, pupils begin to choose what subjects to study, there are many more male teachers (even in girls-only schools) and the staff hierarchy is dominated by men. The oldest pupils are physically mature, and, in the eyes of the youngest pupils, indistinguishable at first sight from adults. There are also certain continuities. School space continues to be dominated by boys (e.g. in the playgrounds, where they play football while the girls stand round watching or talking quietly in groups (Wolpe, 1977)), and the school structures, the formal curriculum and the hidden curriculum all maintain the masculine–feminine divide (Byrne, 1978, and see Chapter 19).

Somewhere in this mass of detail lie important clues, but arguments which focus only on these points fail to take account of another element. If we consider the timing of the break between schools, it may prove to be seriously out of phase with the physiological and emotional development of girls. Girls of this age are usually a year ahead of boys in maturation, and it is only by the age of twelve, when they are well established in secondary school, that the average boy becomes markedly bigger and stronger than the average girl. By contrast, many girls show obvious signs of physiological change, and may even be menstruating before they leave primary school (Lee and Gropper, 1974; Wolpe, 1977). Is it possible that the acceleration in sex differences, and the way the average girl falls behind academically at this stage, is linked to a failure of schools to adjust to the earlier maturing of girls?

Children now mature and interest themselves in sexual matters at an earlier age than ever before. During the first few years of secondary school they are often involved in a commercial culture of adolescence in which pop, cult figures and mass fashions influence their attitudes (Gill, 1977; Sebald, 1968). This commercialism is notorious for its stimulation of romance and glamour along sex-stereotyped lines, and for the fact that it perpetuates the basic divide between masculine and feminine activities. Many boys are obsessed by opportunities to earn money and to prove their 'virility'. So study and the lure of a career can be consistent with male adolescence. For girls there is a potential conflict between the 'distracting presence of sexuality' and the demands of study (Sharpe, 1976). Feminine adolescent culture does not glamorise academic success and careers; instead it promotes an 'ideology of romance, marriage, family life, fashion and beauty' (McRobbie, 1978). Many girls arriving at secondary school are already avid readers of magazines such as *Jackie* which translate sexuality into terms appropriate for their age group, and Wolpe observed eleven-and twelve-year-old girls at school 'behaving in what appeared to me to be an extraordinarily coquettish, feminine manner with particular male teachers. It was a type of behaviour which somehow clashed with their physical immaturity' (Wolpe, 1977, p. 35).

In its most extreme form this feminine adolescent culture takes on specifically anti-school connotations—'To put it briefly: one way in which the girls combat the class-based and oppressive features of the school is to assert their "femaleness", to introduce into the classroom their sexuality and their physical maturity in such a way as to force the teachers to take notice' (McRobbie, 1978, p. 104). By their teens, girls are asserting their own interpretations of femininity, and many adults are taken aback by the gap that seems to yawn between themselves and the next generation.

But how far has this gap been precipitated by the failure of adults to adapt at an early stage? Wolpe suggests that teachers do not recognise the sexual connotations of their own and their pupils' behaviour. Farrell (1978) found that 'sex education' in schools gave more attention to reproductive aspects and the risks of venereal disease than to discussion about the personal and social contexts of sexuality. She also found that parents were extremely embarrassed by the subject. They tended to leave boys to find out for themselves, but felt it important to take some initiative with the girls. It was usually the mother who took this initiative, often around the time of first menstruation, when she would discuss 'where babies come from', menstruation and the 'facts of life' all in the context of motherhood, and in an atmosphere of embarrassed secrecy.

By contrast, many of the pre-capitalist societies studied by anthropologists paid special attention to the need to train maturing adolescents in the personal and social responsibilities connected with sexuality. A girl's first menstruation was often the occasion of ritual which not only marked her readiness for married life but 'included instruction given during these rites, and privately, in sexual intercourse, sex hygiene, and sexual behaviour in general' (Read, 1970, p. 275; see also Richards, 1966). These rituals were not a private affair between an embarrassed mother and girl but involved other female relatives and communal celebration of the fact that another girl had 'become a woman'. Societies such as our own, which have aligned themselves with a puritan ethic, have cast sexuality into a punitive, shameful arena in which mothers have been forced to take action not to celebrate maturity but to prevent their daughters 'making mistakes'.

Few people can talk openly with anyone, let alone their adolescent children, about the 'facts of life', menstruation and contraception (Cousins, 1978). And even today there are relatively few girls who can suspend thoughts of marriage for a time, and experiment with the same freedom as their male contemporaries (McRobbie, 1978). It is still unacceptable for many girls to break through the taboos of traditional sexual and marital codes; instead they have to tread a path which is at times tortuous and risky, proving their sexual attractiveness but avoiding the shame of mistakes. Moreover they move along this path with boys who have absorbed the classic double standard of seeking sexual experience but denouncing girls who willingly participate as 'whores' or 'tarts' (Parker, 1974). If the adults closest to young adolescent girls are ashamed and silent, or even deny their own experience (Konopka, 1966), then the girls turn to each other and try to translate the values they pick up from commercial culture.

Sexuality for many girls is therefore distracting, absorbing and risky—an element which draws them into a feminine and adolescent sub-culture which is not precisely the same as their mothers' but is another route into traditional roles. Pubescent girls have two routes through which they can learn about their sexuality (the adolescent sub-culture and their mothers' protectiveness), but both these hasten 'femininity' and do nothing to counter sex stereotyping.

Schools seldom provide a third route. Many still do not teach 'sex education' at all; those that do, lock girls into the same stereotyped channels. Most parents and teachers are hesitant about who should initiate discussions, and rarely talk to each other about how to approach the 'sex education' of young adolescents. The majority of schools which do actually make some efforts in this direction confine themselves to an almost clinical account of animal and human reproduction, quite separate from discussion of the emotional, personal and social contexts (Farrell, 1978). Even in the late 1970s the majority of lessons in 'sex education' are taught by biology teachers, and this fact must be a signal that schools are in fact maintaining traditional sex structures and logic (Rance, 1978).

Conclusions

This analysis helps to explain why, in each of the IEA countries, girls tend to fall behind boys in academic achievement between the ages of ten and fourteen. Among the sciences biology is a consistent exception—and it is also the only subject that deals at all with the topic of sexuality which is so fascinating and absorbing for young adolescent girls. The culture of femininity may also explain why the girls of lower-status-occupation families should do so much worse than those of professional-status families. Girls of professional parents, and especially those with mothers in career posts, can witness the rewards that follow from prolonged study, and benefit from better communication between their parents and teachers (Douglas, 1967). In addition, if they continue to higher education, they may get 'that crucial space' of years spent at college or university when they have the opportunity to work out their individual approach to adult life and sexuality (McRobbie, 1978).

These interconnections remind us that, in the end, we cannot separate patriarchal logic from the context in which it is being acted out. It is dangerous to assume that notions of masculine and feminine will apply in the same way to every detail of life, whether in Japan, the USA or Italy. Patriarchal logic shapes the continuities within the IEA study, i.e. the fact that, while class differences are highly significant, within each class the average girl achieves markedly less well than the average boy (see Fig. 2.2). But patriarchal logic cannot of itself account for all the differences between the countries. For example, there are significant differences between the average science achievement of girls in Hungary and

Italy—Hungarian girls achieve remarkably high scores, and this must be related to the major drive which went into popularising science at all levels of education. However, within each occupational grouping, they are no closer to the boys than in other countries, which suggests that few efforts were made to 'feminise' science. This finding would be consistent with reports from Eastern Europe which suggest that the post-war period has witnessed the growth of adolescent culture in similar ways to the West (Sebald, 1968). Moreover women who work outside the home are heavily burdened with two jobs, because few political initiatives have been taken to encourage men to share in domestic tasks and parenting (Dodge, 1966).

In Italy the patriarchal continuities are very clear. State policies in relation to women's rights and family life are extremely conservative, and any basic measures such as legalised abortion are thwarted by opposition among the medical and religious authorities. Italy is not as technologically sophisticated as other Western countries and is still weighed down by pockets of appalling poverty and extreme regional variations, so the fact that overall achievement in science is markedly lower is not surprising. Its extreme conservatism in relation to women is revealed by the fact that Italian professional-status girls performed so badly in science that they were outstripped by all Italian boys, even those of the lowest-status groups.

What may seem more surprising at first sight is that the girls of the USA and Sweden should not reveal patterns of achievement which differ from their sisters in less 'liberated' countries. After all, these are the two countries in which legislation against sex discrimination has been established on the most far-reaching basis and conscious policies of positive discrimination have been developed (Forgan, 1978). This has been counteracted by the enormously sophisticated and widespread commercialism connected with adolescence, which encourages pubescent girls to fall off in performance more markedly than elsewhere.

We do not have sufficient evidence as yet to reach firm conclusions. But if my analysis is correct, then girls cannot be expected to achieve unless two forms of action are undertaken to counter 'adolescent femininity'. First, cross-sex activities and interests need to be developed and encouraged from the first days of school, and right through education. Second, parents and teachers need to work together to develop alternative models of adolescent sexuality to which young people can relate. Conscious policies along these lines are essential if female disadvantage in science is to be tackled at root, and if the onslaught of adolescence is to be withstood. Pupils should be given opportunities to discuss maternal and commercial cultures of adolescence, and to develop alternative models of mature, adult sexuality which they can adopt for themselves, on the basis of knowledge, and free from shame and secrecy. If this approach is adopted, then our patriarchal heritage can be shaken to some extent, and female disadvantage in science may be reduced.

References

Barron, R. D., and Norris, G. M. (1976), 'Sexual divisions and the dual labour market', 47–69 in Barker, D. L. and Allen, S. (ed.) *Dependence and Exploitation in Work and Marriage*, Longman.

Belotti, E. G. (1975), *Little Girls*, Writer and Readers Publishing Co-operative.

Blackstone, T. (1976), 'The education of girls today' in Mitchell, J., and Oakley, A. (eds.), *The Rights and Wrongs of Women*', Penguin.

Bujra, J. M. (1978), 'Introduction: female solidarity and the sexual division of labour' in Caplan, P. and Bujra, J. M. (eds.) *Women United, Women Divided: Cross-Cultural Perspectives in Female Solidarity*. Tavistock.

Byrne, E. M. (1978), *Women and Education*, Tavistock.

Children's Rights Workshop (1976), *Sexism in Children's Books: Facts, Figures and Guidelines*, Writers and Readers Publishing Co-operative.

Cohen, G. (1978), 'Women's solidarity and the preservation of privilege', in Caplan, P. and Bujra, J. M. (eds.), *Women United, Women Divided: Cross-Cultural Perspectives in Female Solidarity*. Tavistock.

Cousins, J. (1978), *Make it Happy: What sex is all about*, Virago.

Croll, E. (1978), 'Rural China: segregation to solidarity', in Caplan, P. and Bujra, J. M. (eds.), *Women United, Women Divided: Cross-Cultural Perspectives in Female Solidarity*. Tavistock.

Dodge, N. T. (1966), *Women in the Soviet Economy*, Johns Hopkins University Press.

Douglas, J. W. B. (1967), *The Home and the School*, Panther, Modern Society series.

Farrell, C. (1978), *My Mother Said—The Way Young People Learned about Sex and Birth Control*, Routledge & Kegan Paul.

Forgan, L. (1978), 'Reports on USA', in Women's Page, *Guardian*, 4–6 December.

Fremlin, C. (1979), 'Walking in London at night', *New Society*, April.

Gill, D. (1977), *Illegitimacy, Sexuality and the Status of Women*, Blackwell.

Goldberg, S., and Lewis, M. (1969), 'Play behaviour in the year-old infant: early sex differences', *Child Development*, Vol. 40, pp. 21–31.

Hutt, C. (1972), *Males and Females*, Penguin.

Kelly, A. (1976a), 'Women in physics and physics education', in Lewis, J. (ed.), *New Trends in Physics Teaching*, UNESCO.

—(1976b), 'A discouraging process: how girls are eased out of science', in Hinton, K. (ed.), *Women and Science*, SISCON project.

Kohlberg, L. (1966), 'A cognitive-developmental analysis of children's sex-role concepts and attitudes', in Maccoby, E. E. (ed.), *The Development of Sex Differences*, Stanford University Press.

Konopka, G. (1966), *The Adolescent Girl in Conflict*, Prentice-Hall.

Lee, P. C., and Gropper, N. B. (1974), 'Sex-role culture and educational practice', *Harvard Educational Review*, Vol. 44, 369–410.

Leeds Literature Collective (1973), 'Science and girls; a study of primary science text books', mimeographed; also *Shrew*, Vol. 5, 8 October.

Lewis, M. (1972), 'State as an infant-environment interaction: an analysis of mother–infant behaviour as a function of sex', *Merrill-Palmer Quarterly*, Vol. 18, 95–121.

Lobban, G. (1978), 'The influence of the school on sex-role sterotyping' in Chetwynd, J., and Hartnett, O. (ed.), *The Sex Role System*, Routledge & Kegan Paul.

Maccoby, E. E. (1963), 'Women's intellect', in Faber, S. M. and Wilson, R. H. L. (eds.), *The Potential of Women*, republished in revised form in Hinton, K. (ed.), (1976), *Women and Science*, SISCON project.

Maccoby, E. E., and Jacklin, C. N. (1975), *The Psychology of Sex Differences*, Stanford University Press.

McRobbie, A. (1978), 'Working class girls and the culture of femininity', in Women's Studies Group, (ed.) *Women Take Issue*, Hutchinson.

Mischel, W. (1966), 'A social-learning view of sex differences in behaviour', in Maccoby E. E. (ed.), *The Development of Sex Differences*, Stanford University Press.

Parker, H. (1974), *View from the Boys*, David & Charles.

Rance, S. (1978), 'Going all the way . . . in sex education', *Spare Rib*, No. 75, October.

Read, M. (1970), 'Education in Africa: its pattern and role in social change', in Middleton, J. (ed.), *From Child to Adult: Studies in the Anthropology of Education*, Texas Press and Doubleday.

Richards, A. E. (1966), *Chisungu: a Girl's Initiation Ceremony among the Bemba of Northern Rhodesia*, Faber & Faber.

Saifullah-Khan, V. (1976), 'Purdah in the British situation', in Barker, D. L. and Allen, S. (eds.), *Dependence and Exploitation in Work and Marriage*, Longman.

Sebald, H. (1968), *Adolescence: a Sociological Analysis*, Appleton-Century-Crofts.

Sharpe, S. (1976), *'Just Like a Girl': How Girls Learn to be Women*, Pelican Books.

Shave, S. (1978), '10 ways to counter sexism in a junior school', *Spare Rib*, No. 75, October.

Stacey, J., Béreaud, S., and Daniels, J. (eds.) (1974), *And Jill Came Tumbling After: Sexism in American Education*, Dell Publishing Co.

Torrance, E. P. (1963), 'Changing reactions of pre-adolescent girls to tasks requiring creative scientific thinking', *Journal of Genetic Psychology*, Vol. 102, 217.

Wolpe, A. M. (1977), *Some Processes in Sexist Education*, Explorations in Feminism, No. 1, Women's Research and Resources Centre Publications.

5

Science achievement as an aspect of sex roles

ALISON KELLY

In this chapter I want to consider achievement in school science as a manifestation of sex roles. I will argue that the characteristically low achievement of girls in science is an aspect of the feminine sex role—that is, learned behaviour which is appropriate for females in our society. Sex roles are not totally prescriptive or uniform, and will be followed more or less rigorously in different situations and by different individuals. But few people are unaware of, or completely oblivious to, the prevailing sex roles in their society. So role theory may be able to explain both why most girls under-achieve in science and why a few girls perform satisfactorily.

Theories of socialisation

Theories about the acquisition of sex roles fall into two broad groups—social learning theories and cognitive developmental theories. Social learning theories suggest that parents, peers and teachers play a dominant part in moulding children's behaviour and establishing habits and roles. They can be further differentiated into reinforcement theories and observational theories. Reinforcement theory argues that rewards for sex-appropriate behaviour and punishments for sex-inappropriate behaviour are the main mechanism by which sex roles are acquired. These rewards and punishments may be deliberate and conscious, but equally they may be subtle and unconscious. Because the child is highly dependent on parents' (and teachers') approval, and has a strong need to retain their love, s/he is sensitive to very slight indications of approval or disapproval, such as a glance, a gesture or a tone of voice. The child learns which behaviour is approved and which disapproved and, in order to gain approval, repeats that which has been favoured until it becomes second nature. This position is set out in detail by Mischel

(1966).

But parents and teachers can only reward or punish behaviour which the child already demonstrates. A second aspect of social learning theory is imitation and modelling, and this is particularly important in the acquisition of new behaviour. Observational theory suggests that children will imitate adults and peers (role models) whose behaviour is seen as in some way likely to gain a reward (Bandura, 1971). Children therefore imitate powerful figures (who are seen to gain rewards or who may be in a position, because of their power, to reward those who imitate them). They also imitate people, especially parents, whose approval they seek, and people who are seen doing interesting things. In addition, children imitate people who they see as being like themselves. Same-sex parents, siblings and older children are more similar to a child than opposite-sex models, and observational theory suggests that this is important for the learning of sex roles. The more sex-typed behaviour the child observes the more sex-typed her or his own behaviour and attitudes are likely to become.

Cognitive developmental theories differ from social learning theories in several respects. Cognitive theorists see the child as motivated to achieve competence rather than reward. As part of their attempt to make sense of the world, children develop categorisation rules. Sex is one of the primary categories for people, and being secure in a sex role is one aspect of competence in organising experience. Children put together a cluster of attributes which they label male or female and then they try to copy the appropriate cluster. As Kohlberg (1966, p. 89) puts it, 'the social learning syllogism is "I want rewards, I am rewarded for doing boy things, therefore I want to be a boy". In contrast a cognitive theory assumes this sequence: "I am a boy, therefore I want to do boy things, therefore the opportunity to do boy things (and to gain approval for doing them) is rewarding".' This approach conceives of the child as an active participant in structuring her or his experience, and formulating sex role concepts. Social learning theories see the child as essentially passive, being moulded by external forces; but cognitive theories see the child as essentially self-socialising, first developing categories and then fitting her or himself into these categories.

Cognitive developmental theories can explain some apparently anomalous results. For example, Maccoby and Jacklin (1974, p. 364) quote the case of a four-year-old girl who 'insisted that girls could become nurses but only boys could become doctors. She held to this belief tenaciously even though her own mother was a doctor.' This behaviour is inexplicable with reinforcement or observational theory but fits easily into a cognitive developmental framework: the girl had labelled 'doctor' as 'masculine' (presumably having derived this label from peers and the media) and then insisted that the world should fit her categories. As with other concepts, the child's first ideas about sex roles tend to be crude, concrete, oversimplified and exaggerated. They resemble caricatures and do not allow of exceptions. If the child's conception of 'doctor' is masculine, the fact that women are known and seen to be doctors is immaterial, an anomoly to be rejected rather than assimilated. Only with further development do the concepts become more subtle and sophisticated—teenage children do not make the same mistakes as four-year-olds.

Although these theories are usually presented as competing models of socialisation, it is not, of course, logical to assume that one is right and the others wrong. There is a certain amount of evidence for each of them, and it seems probable that reinforcement, observation and cognitive processes all play a part in children's acquisition of sex roles.

Many feminist accounts of socialisation adopt a crudely behaviouristic approach to child rearing, stressing the differences in the ways girls and boys are treated (e.g. Belotti, 1975; Nicholson, 1977). Certainly there is some research which shows that adults treat male and female children differently from birth and throughout childhood (e.g. Lewis, 1972; Moss, 1974). But Maccoby and Jacklin (1974) summarise their extensive review of the literature as follows: 'our survey of data has revealed a remarkable degree of uniformity in the socialisation of the two sexes' (p. 348). 'Parents seem to treat a child in accordance with their knowledge of his [*sic*] individual temperament, interests and abilities rather than in terms of sex-role stereotypes' (p. 362). Newson *et al.* (1978) illustrate this individualisation of treatment when they report instances of parents who support their children's cross-sex hobbies (knitting or embroidery for boys, football for girls) even when they do not fully approve of them. However, Maccoby and Jacklin's survey did reveal some differences in the ways girls and boys were treated, and they conclude that 'boys seem to have more intense socialisation experiences than girls . . . adults respond as if they find boys more interesting and more attention provoking than girls' (1974, p. 348). Boys received both more praise and more blame from their caretakers, and were handled and played with somewhat more roughly. They were also strongly discouraged from 'sissy' behaviour and any interest in feminine toys or dress. Girls were not, however, particularly discouraged from tomboyishness.

Reinforcement theory demands not only that there be differences in the treatment of boys and girls, but that these differences be causally linked to the acquisition of sex roles. In other words, it implies that children who are treated in a sex-stereotyped way will demonstrate more sex-typing than children who are treated more equally. This link between adult behaviour and child sex-typing is more often assumed than demonstrated. Considering the difficulty of fostering desired behaviour or eliminating undesired behaviour in children by conscious manipulation of rewards and punishments there must be some doubt about the effectiveness of the small, unconscious reinforcements postulated by reinforcement theory. Very little research has been specifically concerned with this point, but McConaghy (1978), in reviewing the existing studies, concluded that 'most found negligible or small effects'.

Feminist authors have also stressed the importance of observational learning, and there is some good evidence for this. For example, mothers who are oriented to life outside the home are more likely to have career-oriented daughters, particulary in non-traditional fields, than more home-bound mothers. On the other hand, mothers who are achievement-oriented for their children but not for themselves produce achievement-oriented sons and daughters who push their own children to succeed (Block *et al.*, 1973; Hoffman, 1972). Nevertheless there are problems with observational theory. As Maccoby and Jacklin say, 'children have not been shown to resemble closely the same sex parents in their behaviour' and 'when offered an opportunity to imitate either a male or female model children (at least those under age 6 or 7) do not characteristically select the model whose sex matches their own . . . yet their behaviour is clearly sex-typed at a much earlier age' (1974, p. 363). Finally, Maccoby and Jacklin point out that much of children's sex-typed behaviour is qualitatively different from adult behaviour—boys but not men avoid the company of females, girls but not women play hopscotch and jacks. It may be that modelling on older children is more important than modelling on adults, but this has received less attention from researchers.

Children's reading books and the media have frequently been shown to be strongly sex-stereotyped (see Chapter 4) and are often held to be a major influence on children's acquisition of sex roles (particularly when the parents are concerned to minimise sex-stereotyping in the home environment). But again the causal link between the sex-stereotyped content of books and media and the children's own sex-typing has not been demonstrated. Such sex-stereotyping can be deplored on the grounds that it is discriminatory and frequently offensive, but we cannot be sure that it directly promotes sex-typing in children.

Cognitive theories are more recent in origin, and although they are gaining increasing support (see Maccoby and Jacklin, 1974; Lockheed and Eckstrom, 1977; McConaghy, 1978) they have not been extensively tested. At first sight it may seem that the fact that children exhibit sex-typed behaviour before they are sure that gender is constant, or can correctly classify other people by gender, presents problems for this theory. But Kohlberg (1966) argues that the process can begin as soon as a child is aware, in any sense, of her or his own gender. Young children often believe that gender depends on hairstyle, dress and behaviour rather than on genitals, and so they conform rigidly to these aspects of sex roles. Kohlberg also argues that the development of sex role concepts is a cognitive process similar to the development of other concepts and shows that the sex role concepts of children who are cognitively advanced (high IQs) develop through the same stages as those of less advanced children, but at an earlier age.

Socialisation theories and science achievement

Having outlined the main theoretical approaches to sex role socialisation, I now want to consider their relevance to sex differences in science achievement. The culture hypothesis that I discussed in Chapter 2 was essentially an application of social learning theories to girls' achievement in science. I suggested there that different societies moulded girls and boys in different ways, and that some of these were more favourable to science than others. According to reinforcement theory this differential moulding would take place through parents, teachers and peers indicating that science is more suitable for boys than for girls. For example, boys might be strongly rewarded for success and strongly criticised for failure in science while the reaction to girls' performance was more neutral or even hostile (see Chapter 17). Reinforcement theory would also suggest that parental approval and disapproval are important in ensuring that boys and girls play with different sorts of toys. This will affect performance in science because boys' toys (such as mechanical building kits, electric trains and chemistry sets) are more efficacious in developing scientific skills than are girls' toys (such as dolls, prams and teasets). Boys' hobbies and household tasks also have a more technical orientation than girls'. Judy Samuel (Chapter 18) presents some evidence for this when she notes that first-year boys seem more at home in the science laboratory and more familiar with the equipment than first-year girls. Observational theory would suggest that the sex difference in science achievement is produced by observation and role modelling. Children notice that an overwhelming majority of scientists and science teachers are men and that science textbooks contain many more illustrations of and references to boys than girls (see Chapters 4 and 18). They copy the models which are available to them.

While not wishing to deny the importance of these factors in shaping girls' attitudes to science, I am not convinced that they are sufficient to account for the uniformity of the observed sex differences in achievement from country to country. As I argued in Chapter 2, the degree to which differential reinforcement for science achievement is given to girls and boys probably varies between cultures and sub-cultures. Certainly the availability of female role models in science varies. Yet the sex differences in achievement do not vary to any great extent. Girls do not do better in science in schools where a larger proportion of science teachers are women, nor in countries where a larger proportion of science students are women. The gap between the sexes in science is no smaller in countries with an active commitment to equality between the sexes than in countries with no such pretensions. Elinor Kelly has argued in Chapter 4 that these differences between cultures are minor compared to the overwhelming similarity of their patriarchal cultures and the constraints placed upon girls at the time of puberty. Again I would not wish to deny these similarities. But I maintain that the differences between the cultures are real as well. Sharpe (1976, p. 148), in discussing girls who reject mathematics and science because they see them as being too difficult and technical for girls to understand, says, 'the influence of culture on this attitude is indicated by the variation which exists between different nations in the extent to which girls participate in science and mathematics courses'. I agree. Nations do vary in their cultures, and this affects the girls' attitudes to and participation in science. But by the same token the lack of variation between different nations in the extent of the achievement gap between the sexes in science suggests that culture is not the determining factor in science achievement.

The cognitive developmental theory can, however, explain the constancy of the sex differences in science achievement from country to country. If science achievement has a masculine image in any society, then boys will be motivated to achieve competence in science as part of their developing masculinity; conversely girls will see success in science as incompatible with their developing femininity and so avoid it. The overall impression of science as masculine (which is common to all the cultures considered in the IEA study) is more important than how strongly this impression is reinforced in any particular culture. The common image in all countries produces a common sex difference in all countries.

It is easy to understand how science acquires a masculine image in children's eyes. In all the countries studied the majority of scientists are men; textbooks and the media frequently represent science as almost exclusively a male preserve. So children can attach a masculine label to science even if they are not specifically encouraged to do so, and even if their immediate surroundings do not generate such a label. In Chapter 16 Helen Weinreich-Haste shows that, at least in England, physical science does indeed have a strongly masculine image. By contrast, biology has a fairly neutral image, which accords with the finding that sex differences in biology achievement are small.

Cognitive theories and social learning theories draw upon much of the same data. Feminists can argue from either position that women's exclusion from science is a manifestation of their subordinate position in patriarchal society, and hence of their exclusion from powerful positions and powerful knowledge in that society. But patriarchy is not of itself an explanatory concept, and the theories differ in their accounts of how it operates. In particular, they differ in the roles they assign to the child and the society. On the cognitive developmental theory the child is the active socialising agent; society merely presents an image of masculine and feminine to the child. By contrast, social learning theories hold that society actively exerts pressures on the child to conform to these images. Whereas social learning theorists would see children's sex-typed toy preferences as the outcome of parental sex discrimination, cognitive theorists would suggest that children are self-motivated to choose sex-typed toys (provided they can pick up some hints, from any source, as to what is appropriate and what is inappropriate for their sex). There would be no disagreement, however, over the suggestion that 'boys' toys' are more effective than 'girls' toys' in developing science-related skills and knowledge.

Cognitive theory does have certain advantages over social learning theory. It can explain the puzzlement and disappointment that many parents who try to raise their children in a non-sexist way feel over the behaviour of their four-year-olds. It can also explain how some individuals brought up in a traditional way come to question their own sex stereotpyes, since concept formation is seen as an active process which changes as the individual develops. Children's attitudes towards sex roles generally become less stereotyped as they mature (Vestin, 1975), and this too presents problems for social learning theory. Both reinforcement and observation would be expected to produce increasingly strong sex typing as they continue to operate throughout childhood, but in fact sex typing is frequently very strong among young children and then declines.

The IEA study did, however, show that the gap between boys' and girls' science achievement widened between the ages of ten and fourteen, a time when sex role concepts should be becoming more mature. This is also a time of compulsory schooling and often of compulsory science. The results reported in Chapter 2 did not support the school hypothesis that girls and boys benefit from different types of science teaching. They fit more satisfactorily into a feedback model, on the lines of 'to them that have shall be given'. In technological societies boys probably have extensive background experiences in science, from their play activities and household tasks, before they begin to study it formally in school. They are ahead of girls in science achievement at ten years old, and consolidate this lead to be even further ahead by fourteen years old. It looks as though science teaching builds upon foundations of incidental learning in science which boys possess but which girls lack. Given the cumulative nature of school science, minor uncertainties in the initial stages may generate major incomprehensions later on. In addition a feedback loop between attitudes and achievement seems almost inevitable. High-achieving pupils will be positively reinforced by good marks, and so develop good attitudes towards science; pupils with positive attitudes towards science will be motivated to achieve well. So if boys begin the study of science with better attitudes than girls towards science (because of its gender connotations or because of their greater experience in this field), the attitude–achievement feedback loop will act to increase their lead in both attitudes and achievement. There may also be a resurgence of interest in, and conformity to, sex roles around puberty (i.e. between the ages of ten and fourteen) when children again become concerned with the sex appropriateness of their behaviour, including success in science.

The implications of socialisation theories

I have discussed socialisation theories and the differences between these theories in some detail because I think it is important to understand *how* sex-typing takes place if we wish to counter it. The different theories have rather different implications for action. The implications of social learning theories are fairly straightforward—reward children for behaviour traditionally considered inappropriate for their sex as well as for behaviour traditionally considered appropriate; provide role models of men and women engaged in cross-sex-typed activities; put pressure on book publishers, television companies and schools to eliminate sex stereotypes from the material they publish and use. In other words, try to create an androgynous world in which differentation between the sexes is eliminated or reduced to a minimum.

Cognitive theories suggest that this strategy may be ineffective or even counter-productive in the short term. As McConaghy (1978) put it, 'if cognitive theory is correct in assuming gender to be important to young children and to be understood by them as determined by sex-role behaviour, then discouragement of sex-role learning unaccompanied by clarification of the nature of gender may result in rigid conformity to sex-role norms, and may even arouse anxiety in the child over gender identity'. In other words, to attempt to eliminate sex distinction may make children more conformist to those distinctions they can establish, and may even cause them anxiety over their gender roles. McConaghy suggests that children should be brought to realise as soon as possible that gender depends upon genital formation rather than upon behaviour. They should also be encouraged to see cross-sex-typed behaviour as not only pleasing to the adults around them but as competent behaviour for their own sex. This suggests that, in the short term at least, we should be aiming not towards a neutral or sex-free conception of behaviour, but towards reversed or dual-sex typing.

With respect to science this means that we should be aiming towards giving science a more feminine image and convincing young girls that achievement in science is a normal and acceptable part of feminine behaviour. Some ways in which this might be attempted are suggested in the final chapter of this book. These remedies may take some time to be effective. If the roots of girls' underachievement in science are located in an area as diffuse as the image of science, then swift and decisive action is impossible. But this is not a council of despair. The fact that the image of science is diffuse means that it involves everyone, and conversely everyone can be involved in changing it. There is no need to wait for curriculum developers to produce new materials. Every teacher and every parent can examine their own stereotypes and take steps to alter the image of science encountered by their children.

Other theories

There are both similarities and differences between my position and those of the other authors in this section. Although I do not reject the possibility, or the possibility of demonstrating, a biological basis to the sex differences in science achievement, I do not find the present evidence compelling. As Jeffrey Gray himself admits (Chapter 3), 'there is a formidable array of negative evidence' for the X-linked hypothesis. But even if a biological contribution to girls' under-achievement in science were to be established, I do not think that would be the end of the story. Biology is *not* destiny. Society has the option, through schooling and socialisation, of providing additional training in the areas where each individual is weakest so as to produce citizens with well rounded personalities and competences. Biological factors provide only a predisposition to acquire certain skills, and do not imply that those skills are unlearned or unlearnable. I do not agree with Gray that we should 'welcome the different patterns of intellectual abilities shown by the two sexes'. Any limitation upon an individual's capabilities is a deficiency which deprives her or him of some experience or opportunity. Neither do I agree that women are not disadvantaged by their virtual exclusion from science; I have suggested in Chapter 1 that they *are* disadvantaged, both individually and as a social group. The argument about top jobs and the recognition of outstanding talents is a red herring: for the bulk of the population technical jobs requiring some measure of scientific education are better rewarded than non-technical jobs; the fact that one outstanding woman can make her way through the system by hard work, determination and brilliance does not imply that the system is open on equal terms to all men and women. As Maccoby and Jacklin (1974, p. 367) point out, there is an enormous overlap between the sexes in all measured abilities, so that even if the requisite skills were completely genetically determined there would be 'more than enough women to fill our engineering schools if women's talents were developed through the requisite early training and interests'.

My disagreements with Elinor Kelly (Chapter 4) and Esther

Saraga and Dorothy Griffiths (Chapter 6) are more on matters of emphasis than substance. Elinor Kelly's account of socialisation in a patriarchal society is compelling, but, as I have already indicated, I do not think that by themselves social learning theories of the acquisition of sex roles are adequate. Socialisation is not a crude matter of repressing young girls' natural curiosity and independence. It is a subtle and delicate process in which the developing child plays an active part. Saraga and Griffiths are more concerned with the political factors underlying women's exclusion from science than with the means by which this exclusion is maintained—the why rather than the how. They suggest that 'socialisation can . . . appear to be arbitrary, and open to change through little more than a campaign to change attitudes'. But neither Elinor Kelly nor I would see socialisation as arbitrary—for both of us it is deeply rooted in the patriarchal nature of our society. The real disagreement is over the status of a campaign to change attitudes. Whereas Saraga and Griffiths seem to see this as a fairly minor matter I would argue that sexist attitudes are fundamental to patriarchy, and that a campaign to change attitudes can be truly revolutionary. I also disagree with Saraga and Griffiths's contention that the *subject matter* of science is determined by industrial and military requirements. These factors clearly influence the applications of science, but the content of school science is more autonomous. It reflects the logical, conceptual and mathematical structure of science as much as its applications. Indeed, employers frequently complain that school science is insufficiently influenced by industrial requirements.[1]

Note

1 I would like to thank Helen Weinreich-Haste, Elinor Kelly and Judy Samuel for reading and commenting on an earlier draft of this chapter.

References

Bandura, A. (1971), 'Analysis of modelling processes', in Bandura, A. (ed.), *Psychological Modelling: Conflicting Theories*, Chicago, Aldine-Atherton.

Belotti, E. G. (1975), *Little Girls*, Writers and Readers Publishing Co-operative.

Block, J. A., Von der Lippe, A., and Block, J. H. (1973), 'Sex role and socialization patterns: some personality concomitants and environmental antecedents', *J. Consulting and Clinical Psychology*, Vol. 41, 321.

Hoffman, L. W. (1972), 'Early childhood experiences and women's achievement motives', *J. Social Issues*, Vol. 28, 129.

Kohlberg, L. (1966), 'A cognitive-developmental analysis of children's sex-role concepts and attitudes', in Maccoby, E. E. (ed.), *The Development of Sex Differences*, Stanford University Press.

Lewis, M. (1972), 'Parents and children: sex-role development', *School Review*, Vol. 80, 228.

Lockhead, M. E., and Ekstrom, R. B. (1977), *Sex Discrimination in Education: a Literature Review and Bibliography*, Educational Testing Service, Princeton, New Jersey, Research Bulletin, RB–77–5, May.

Maccoby, E. E., and Jacklin, C. N. (1974), *The Psychology of Sex Differences*, Stanford University Press.

McConaghy, M. J. (1978), 'Determinants of children's sex-role learning', Paper presented at World Congress of Sociology, Uppsala, Sweden, August 1978.

Mischel, W. (1966), 'A social-learning view of sex differences in behaviour', in Maccoby, E. E. (ed.), *The Development of Sex Differences*, Stanford University Press.

Moss, H. A. (1974), 'Sex, age and state as determinants of mother-infant interaction', in Stone, L. J., Smith, H. T., and Murphy, L. B. (eds.), *The Competent Infant: Research and Commentary*, Tavistock.

Newson, J., Newson, E., Richardson, D., and Scaife, J. (1978), 'Perspectives in sex-role stereotyping', in Chetwynd, J., and Hartnett, O. (eds.), *The Sex Role System*, Routledge & Kegan Paul.

Nicholson, J. (1977), *What Society does to Girls*, Virago.

Sharpe, S. (1976), *'Just Like a Girl': How Girls Learn to be Women*, Penguin.

Vestin, M. (1975), *Some Information concerning the Sex Role Question and Programs for Development of Equality between Men and Women specially within School and Education*, Swedish Board of Education.

6

Biological inevitabilities or political choices? The future for girls in science

ESTHER SARAGA and DOROTHY GRIFFITHS

In this section of the book a number of competing theoretical frameworks are proposed to account for the IEA survey data reported by Alison Kelly. The most important feature of the framework we shall use is that, unlike the others, we do not wish to emphasise one particular factor (be it biological, psychological, sociological or educational). We believe that the relationship of girls to science, and their performance in it, are too complex to be understood in terms of one factor, but that several factors must be integrated in a broader understanding of the social context in which science is carried out, and in which socialisation takes place. More specifically, we shall argue that theories couched in biological terms cannot be sustained; that socialisation theories do not go far enough because they do not attempt to explain *why* socialisation follows particular patterns; and that it is not sufficient just to consider the development of girls in relation to science—the development and practice of science itself must also be discussed.

Despite social, economic and political differences, all the countries in the IEA survey are divided by sex. Socialisation takes place within, and is structured by, a social order in which, despite the economic indispensability of their domestic labour, women are subordinated to men and are largely excluded from positions of power and prestige. Science too, is stratified, since some disciplines, especially the physical sciences, are more relevant and essential to the industrial and military power of these societies. Given the exclusion of women from powerful and prestigious positions, we would predict, not only that girls will perform less well than boys in science, but also that they will be less likely to enter the physical than the biological sciences.

Improving the position of girls in science is important to most of the other contributors to this volume, but their failure to acknowledge the complexity of the problem and the importance of locating it in its socio-historical context may have important consequence for the development of strategies for change. In particular it may encourage false optimism if they believe that socialisation patterns can be fairly easily altered; or, alternatively, the failure to achieve significant changes may lead to false pessimism, and the countenancing of biological explanations.

At present, biological explanations of sex differences, which suggest severe limits to change, have appeared frequently, both in psychological literature (for example, Hutt, 1972, Buffery and Gray, 1972; Coltheart, 1975; Newcombe and Ratcliffe, 1978) and in the more popular media (for example, Eysenck in *Vogue* 1978; Bellamy in *Radio Times*, 1978). The mystique of science, and the consequent deference to scientific expertise, make biological explanations particularly powerful. We will therefore begin by outlining our objections to these explanations. We will then summarise our points of agreement and disagreement with the socialisation theories, and finally move to a fuller account of our own ideas.

Is female under-achievement biological?

The consistency of sex differences in test scores across countries might at first glance appear to suggest that the differences are innate. In this volume, this position is represented by Gray, (Chapter 3) who argues that visuo-spatial ability is '. . . strongly involved in much scientific thinking . . .'; and there is certainly evidence that, at least after puberty, boys perform better than girls on a variety of tests claimed to measure visuo-spatial ability (see, for example, Maccoby and Jacklin, 1975).

Gray devotes most of his chapter to an examination of whether this sex difference can be accounted for by a specific genetic hypothesis, concluding that '. . . at the moment . . . the status of the X-linked hypothesis is uncertain'. Notwithstanding this, he argues that 'there are a number of reasons . . . to suppose that some kind of biologically based account of the sex difference in spatial ability will eventually turn out to be correct'. His arguments in support of this contention, however, are open to challenge on several grounds.

First, he does little to substantiate his claim that spatial ability is strongly involved in scientific thinking, nor does he justify the importance of purported differences for different sciences. He says merely that '. . . *intuitively* one would expect the requirement for spatial thinking to be greatest for physics and least for biology, with chemistry somewhere in between' (our emphasis). He thus appears to assume that each discipline is homogeneous in the cognitive skills it requires. Clearly this needs further research, but it seems unlikely that, say, evolutionary and molecular biology involve identical cognitive skills. Certainly in physics many problems can be solved in either spatial or mathematical terms. Furthermore, how children approach a scientific problem is likely to depend upon the way they have been taught, and it is significant that the IEA survey found that traditions in approach to science teaching varied markedly in different countries. Comber and Keeves (1973, p. 95) comment that '. . . in the English-speaking nations there is a marked empirical approach to science teaching, with an emphasis on laboratory work and practical experience. The Continental European nations would appear to have built up a strong theoretical tradition with less emphasis on investigation and inquiry. Japan has developed independently a pattern strikingly its own.' These differences in style might not only account for some of the striking differences in test scores between countries (an aspect of the data that Gray ignores) but also indicate that the link between 'ability' and performance in science is rather more complex than Gray suggests.

It is also surprising that he focuses almost exclusively on the link between visuo-spatial ability and science, thus ignoring the importance of mathematics, particularly for physics: especially as the studies of spatial ability that he cites predict success in various mathematical skills. We cannot help wondering whether consideration of mathematics causes difficulty for his approach, since sex differences in mathematics do not appear to be as great as in physics (see Table 1.2), and, moreover, no specific biological model of mathematical ability has been proposed.

His emphasis on abilities ignores the importance of a whole range of factors—anxiety, expectations, motivation, girls' negative attitudes to science, etc.—which are known to influence test performance (see Chapters 2, 7 and 16; also Torrance, 1963). Maccoby has suggested that the development of spatial ability itself may be socially influenced; girls who perform well on tests of spatial ability are more likely to have been encouraged by their mothers to be assertive and independent (traditionally masculine traits) (Maccoby, 1970). This is supported by cross-cultural studies which demonstrate that where greater independence is stressed in girls' socialisation there is no sex difference in spatial ability scores (Berry, 1966; MacArthur, 1967).

A convinced biological determinist could argue, of course, that these emotional and temperamental differences are themselves biologically based, and that cross-cultural differences derive from genetic differences between 'races'. However, such claims would have to be substantiated. Biologically based theories of sex differences in emotional behaviour are subject to many of the criticisms that we have made about theories concerning spatial ability. And genetic theories about 'racial' differences are based on a confusion between the social and biological uses of the term 'race'. While, socially, races are defined in terms of physically observable characteristics, biologically, variation *between* races is much less than variation *within* one race (see, for example, Rose, 1976).

Our second criticism concerns Gray's claim that the existence of a similar sex difference in animals is 'presumptive evidence that biological factors are indeed at work' in humans. He cites evidence that male rats 'are consistently superior at solving complex mazes'. He thus assumes that the same ability is being measured in rats and humans, even though he later suggests that different 'human' tests may not even involve the same ability. He further argues that in rats the sex difference in spatial ability depends on the presence of testosterone in the neo-natal period, which, despite several crucial qualifications, 'raises the possibility that the human sex difference . . . is similarly dependent on early hormonal influence'. Such an argument is both simplistic and mechanistic, for, as Lambert (1978) has argued, 'At present science has a very inexact conception of how genes, hormones or external inputs affect the biological basis of behaviour at the cellular and molecular levels, but it is certain that *both intrinsic and extrinsic factors* are important' (our emphasis).

Assuming that legitimate conclusions about human social behaviour can be derived from studies of animal behaviour is, of course, one of the fundamental tenets of the biological determinist's position. But it ignores important discontinuities between humans and other animals (for example, the significance of language and cultural transmission, the enormous capacity for learning, the capacity to change and create new environments, and the division between mental and manual labour), which render such extrapolations dubious. The related assumption of evolutionary determinism—that contemporary human social behaviour is a manifestation of traits fixed during evolution—is based on very little actual knowledge of the social structure and behaviour of humans in hunter-gatherer societies, and rests on *post hoc* links between particular traits and social roles (for example, spatial ability and hunting). (For a fuller discussion and criticisms of these kinds of arguments see Archer, 1976, 1978; Slater, 1977; Lambert, 1978; Griffiths and Saraga, 1979.)

Gray's challenge to 'environmentalists' to produce a model as specific as the X-linked hypothesis is one that we shall ignore, not just because we do not consider ourselves 'environmentalists' but also because we do not believe that the complexity of human behaviour and social organisation can stand reduction to a simple determinist model.

The importance of socialisation

Alternatives to biologically based theories are usually posed in terms of 'socialisation' or 'conditioning'. For many people (psychologists and non-psychologists alike) this is a process which involves only external influences; that is, girls and boys are treated differently. However, as Alison Kelly suggests, (Chapter 5) 'self-socialisation' is also important: children observe and construct their own rules about the world and consequently learn at a very early age, even with dedicated 'non-sexist' parents, that certain activities, including science, are considered to be more appropriate for one sex. Elinor Kelly (Chapter 4) demonstrates the necessity to consider socialisation within the framework of a patriarchal society, and stresses the significance of gender identity. The importance of these arguments is that girls growing up in a patriarchal society will internalise beliefs, attitudes and expectations about science, themselves and their future roles which will generate not only negative attitudes towards activities such as science—which are seen as male-appropriate—but which also press against the choice of cross-sex activities and result in female under-achievement in them. Many of the chapters of this book illustrate these processes clearly, and the importance of socialisation must be recognised in any theoretical framework which attempts to explain the IEA data. However, much of this work is purely descriptive, and does not examine the basis of particular patterns of development. Socialisation can therefore appear to be arbitrary, and open to change through little more than a campaign to change attitudes. Male domination and socialisation are not arbitrary, though; they relate to women's position in society, particularly in the family, and they should be examined and explained by reference to the underlying economic, social and political forces which structure them. We move now to a tentative attempt to establish the relevant parameters of such an explanation.

Women's position in society

In all the 'developed' countries studied in the IEA survey, regardless of differences in formal political ideology and the organisation of production, there is a clear social division of labour between the sexes. Men predominate in positions of power and influence, while women occupy low-status jobs and work as housewives and mothers. While men are defined in terms of their work outside the home, women are defined in relation to the family. Thus in advanced industrial societies women have a dual role: they are responsible for the production and maintenance of the labour force, at the same time as constituting a part of it. And this dual role means that girls and women are subject to a set of demands for which there is no parallel for men.

The countries in the sample differ both in the range of education and employment opportunities for women, and in the extent to which women have control over their child-bearing role through the provision of contraception, abortion and nurseries. However, in none of them, including Hungary, where the greatest social change has taken place, has women's position in the family been challenged. Simply considering the country's political ideology (as Alison Kelly does in Chapter 2) is insufficient, because ideologies of equality do not necessarily generate equality. Moreover egalitarian ideologies do not develop in a vacuum but are related to a whole variety of other factors which may effect women's participation in the labour force, especially the relative availability of male labour.

Until the structural basis of women's oppression—in particular, the family—is challenged and changed, women will be unable to take advantage of 'equal' opportunities. They will always be disadvantaged, compared to men, while they carry the prime responsibility for domestic labour and child care. Even if it were possible, as Alison Kelly suggests (Chapters 5 and 20) to change the image of science to make it more attractive to girls, women's feelings of subordination, inferiority and (we suspect) their low status will remain unless there are concomitant changes in their structural position.

In developing a theoretical framework which accommodates the data, therefore, we have to understand the development of girls and boys in a society premised on sexual inequality, for socialisation processes arise from, and reproduce, the present sexual division of labour. However, very little research has been undertaken on the questions of how we develop our self-identity in the context of a particular society, how biological factors such as menstruation and pregnancy are socially mediated, or how gender is reproduced across generations. (See, for example, Rubin, 1975, for recent feminist discussions on these questions.)

Most socialisation theories completely fail to consider the meaning attached to being female or male in a particular society, and assume furthermore that the processes of learning to be feminine or masculine are parallel. There are important differences, though. For example, children of both sexes are generally brought up by women, at least for the first few years of life, and this is likely to have a different significance for girls and boys. Thus it has been suggested that boys have to break from their initial identification with their mother and 'solve the problem' of what 'masculinity' means, while for girls early development is more continous, and 'feminity' can be eaily understood in terms of motherhood (Chodorow, 1974; Lynn, 1962). In Britain and elsewhere early schooling is usually provided by women, and boys and girls learn that as they grow up their care and education will increasingly be the responsibility of men. There is greater pressure on boys to conform to their sex role at an early age, as is illustrated both by the different connotations of 'tomboy' and 'sissy', and by boys being less likely to display cross-sex interests. Consequently children learn at an early age about the higher status of men and masculine activities.

Thus we are arguing that, in terms of women's position, the similarities between the countries in the sample are much greater than the differences. Women's work within the family is of such social and economic importance that it will require much more than a change of attitude to alter, significantly, their opportunities. In education and employment women can never be equal to men as long as the sexual division within the family, and the family itself, remain unchallenged. Apart from some challenges in the immediate post-revolutionary period in the USSR, the position of women and the role of the family have not featured on the agenda for change in Eastern European States; while in Britain, for example, the leaders of both the main political parties are still stressing the importance of the family.

Science is male-dominated

We have argued thus far that the consistent under-achievement of girls in science, compared to boys, across a range of countries is a consequence of the subordinate position of women in all of them. It is often argued, however, that science is not simply male-dominated but that it is in some sense 'masculine', and it is important to consider both what this means, and what it implies, for changes in the position of women in science. If science is masculine, then, can a more feminine science be created?

Obviously science is masculine in the sense that men are numerically predominant. But the idea that science is masculine is usually taken to mean rather more than this. In particular it is suggested that the personality traits characteristic of successful scientists are those which are stereotypically male. As we and others in this volume have already suggested, this personality incompatibility is an important factor operating against girls choosing a cross-sex activity like science, especially as the choice has to be made at a time when they are particularly conscious of their gender and anxious to conform to gender roles. (We recognise that not all girls experience this; some want to do 'male' subjects but experience these kinds of attitudes from their teachers—see Chapter 17, and *Spare Rib*, No. 75, 1978—while others make the choice without experiencing any tension.)

Related to this is the issue of subject choice within science; different sciences, it is suggested, have different images, with the physical sciences being seen as more masculine than the biological sciences. The subject matter of physics appears impersonal, inanimate, and far removed from the world of everyday objects and people. Biology, on the other hand, with its concern for living things, appears more personal and alive, and closer to the everyday world of values and emotion, which women are expected to inhabit. Choosing the biological as opposed to the physical sciences thus involves girls in fewer contradictions, and they receive more encouragement and support in their choice.

There is certainly some evidence (see, for example, Chapter 7 and Chapter 16) that different sciences are viewed as more or less masculine. But there is, at the same time, the danger that much of the characterisation of activities as people- or object-oriented is *post hoc*: activities that girls are interested in becoming by definition 'concerned with people'. For example, even though Newson, *et al.* (1978) point out that girls' proficiency in the use of sewing machines is never treated as evidence of mechanical aptitude, they nevertheless also suggest that girls may see sewing machines as a mechanical aid 'to a person-oriented goal' (making something for someone).

If the position of girls in science is to be improved it is important to consider whether these characteristics are inherent in the subject matter of science. Alison Kelly clearly does not accept this, for she proposes efforts to 'feminise' science by changing the direction of its subject matter, for example via links with traditional female concerns such as the home, the body, and the community (Chapter 20). While we agree that science is not inherently masculine, we are not optimistic about these proposals, since the subject matter of science is not arbitrary. But neither is it just a consequence of the working out of an inherent and internal logic, nor is it simply structured by male domination. Science develops in relation to specific social and historical circumstances, and in order to understand why girls enter biology rather than physics we need to consider both the social role of science in general and the particular roles of the different sciences.

In advanced industrial societies, both capitalist and socialist, science is developed in relation to two major objectives. First, innovation is integral to increasing the efficiency of production through product and process improvement and change; second, it is integral to the development of means of social control. (For a fuller exposition of these ideas see Griffiths *et al.*, 1979.) Simply expressed, science develops in the service of the dominant interests in any society, both strengthening and defending them. The status of the different sciences, therefore, will vary according to their perceived economic and military significance at any particular time. And the higher their status, then the greater the exclusion of women in male-dominated societies, and the less relevant the subject matter is to issues sanctioned as legitimate for female concern. Very little research on the history of science has examined the different sciences from the perspective of their relative status and the role women have played in them. The work which has been done suggests that such research could help to account for the development of the images of particular sciences (see, for example, Rossiter, 1974; Kohlstedt, 1978).

Thus the variation in male domination between, say, physics and biology follows from their different historical, economic and military significance, rather than from some more nebulous 'degree of masculinity'. Science is both a social product and a social process; changing its subject matter to render it more attractive to girls will not be easy, for, if the subject matter reflects the interests and needs of dominant groups, then it is those interests and needs which must first be changed. And, as we have pointed out, such a change will involve a fairly major social transformation.

Conclusion

In this chapter we have suggested that the consistent under-achievement of girls in science, and their relatively poorer performance in the physical as compared with the biological sciences, can only be understood by reference to a number of factors; and we have begun to indicate the kind of theoretical framework which we feel accommodates the IEA data most appropriately. More work is clearly needed on this, but we believe that an appropriate framework should be based on an understanding of each of the following: first, the structural bases of women's oppression; second, the consequent differences between the socialisation experiences of girls and boys; and, third, the industrial and military significance of science, particularly physical science, in the twentieth century.

In order to compare the results of the different countries more information is needed about science education, and its relative availability to girls and boys. In Britain, concern has been expressed about the manpower shortage in science and technology since the end of the last war (for example, the Barlow Report, 1946; the Zuckerman Report, 1961; the Dainton Report, 1968; the Swann Report, 1968), and we have lived through the white heat of Harold Wilson's technological revolution, yet girls and boys have not received equal educational provision in science. In the white heat of the 1960s, for example, there was an enormous increase in expenditure on science by both industry and the State, and a string of new 'technological' universities were created. Yet government reports at the time were emphasising the importance of different curricula for girls and boys (for example, Newsom, 1963). And we believe that much of the current concern about the position of girls and women in science and engineering stems more from a concern about a shortage of skilled personnel than from a concern about inequality. (For example, 'Are women the answer to engineering manpower shortage?', *IEE News*, December 1977, or '. . . the greatest waste of technical potential in this country': evidence submitted to the Finniston Committee by the Women's Engineering Society.)

In dealing only with the performance of fourteen-year-olds the IEA survey results reported here do not direct us to consider why girls and women drop out of science more rapidly than boys. Science as a social practice is a microcosm, in many ways, of the wider society, and women who enter it are faced with a number of problems, discrimination apart, which men do not experience. Promotion in science depends, partly at least, on a system of patronage and sponsorship, and women find themselves excluded from both the formal and informal networks. They find it hard to get taken seriously, their work is undervalued, and they find it difficult to combine a career in science with their domestic responsibilities, especially child care, for women scientists, unfortunately, cannot have wives (White, 1970; Women and Science Collective, 1975; Griffiths, 1978).

It should be clear that we do not believe there is an easy 'solution' to the 'problem' of girls and science. The position of girls, in our view, cannot be accounted for in terms of either biology or socialisation but only as a product of *political* choices. And it can, therefore, only be changed by political means. While not denying the potential for short-term improvements, we believe that the position of girls and women in science will only be fundamentally changed through a political challenge to the structural bases of their inequality.[1]

Note

1 We would like to thank Richard Compton, John Irvine, Ray Holland and Alison Kelly for their helpful comments on this chapter.

References

Archer, J. (1976), 'Biological explanations of psychological sex differences', in *Exploring Sex Differences*, ed. B. B. Lloyd and J. Archer, London: Academic Press.

—(1978), 'Biological explanations of sex role sterotypes', in *The Sex Role System*, ed. J. Chetwynd and O. Hartnett, London: Routledge & Kegan Paul.

Barlow Report (1946), *Scientific Manpower*: Report of the Committee appointed by the Lord President of the Council, Cmnd 6824, London: HMSO.

Bellamy, G. (1978), 'Going to the chair', *Radio Times* 2–8 September, 9.

Berry, J. W. (1966), 'Temne and Eskimo perceptual skills', *International Journal of Psychol.*, 1, 207–29.

Buffery, A. W. H., and Gray, J. A. (1972), 'Sex differences in the development of spatial and linguistic skills', in *Gender Differences: their Ontogeny and Significance*, ed. C. Ounsted and D. C. Taylor, Edinburgh and London: Churchill Livingstone.

Chodorow, N. (1974), 'Family structure and feminine personality', in *Woman in Sexist Society*, ed. V. Gornick and D. Moran, New York: Basic Books.

Coltheart, M. (1975), 'Sex and learning differences', *New Behaviour*, 1, 54–7.

Comber, L. C., and Keeves, J. P. (1973), *Science Education in Nineteen Countries*, London: Wiley.

Dainton Report (1968), *Enquiry into the Flow of Candidates in Science and Technology into Higher Education*, Cmnd 3541, London, HMSO.

Eysenck, H. (1978), 'Why can't a woman be like a man?', *Vogue*, March, 78–9.

Finniston Report (1980), *Engineering Our Future*, Cmnd 7794, London, HMSO.

Griffiths, D. (1978), 'Hell for a woman', *Icon*, 14 (or available in mineograph from the author).

Griffiths, D., and Saraga, E. (1979), 'Sex differences in cognitive abilities: a sterile field of enquiry?', in *Sex Role Stereotyping*, ed. O. Hartnett, G. Boden and M. Fuller, London: Tavistock.

Griffiths, D., Irvine, J., and Miles, I. (1979), 'Social statistics: towards a radical science', in J. Irvine, I. Miles and J. Evans (eds.) *Demystifying Social Statistics*, London: Pluto Press.

Hutt, C. (1972), *Males and Females*, Harmondsworth: Penguin.

Kohlstedt, S. G. (1978), 'In from the periphery: American women in science', 1830–80', *Signs*, 4, 81–96.

Lambert, H. H. (1978), 'Biology and equality: a perspective on sex differences', *Signs*, 4, 97–117.

Lynn, D. B. (1962), 'Sex role and parental identification', *Child Development*, 33, 555–64.

MacArthur, R. (1967), 'Sex differences in field dependence for the Eskimo', *International J. of Psychol.*, 2, 139–40.

Maccoby, E. E. (1970), 'Feminine intellect and the demands of science', *Impact of Science on Society*, XX, 13–28.

Maccoby, E. E., and Jacklin, C. N. (1975), *The Psychology of Sex Differences*, London: Oxford University Press.

Newcombe, F., and Ratcliffe, G. (1978), 'The female brain: a neuropsychological viewpoint', in *Defining Females*, ed. S. Ardener, London: Croom Helm.

Newsom, J. (1963), *Half our Future: a report of the Central Council for Education (England)*, London: HMSO.

Newson, J., Newson, E., Richardson, D., and Scaife, J. (1978), 'Perspectives in sex role sterotyping', in *The Sex Role System*, ed. J. Chetwynd and O. Hartnett, London: Routledge & Kegan Paul.

Rose, S. (1976), 'Scientific racism and ideology: the IQ racket from Galton to Jensen', in *The Political Economy of Science*, ed. H. Rose and S. Rose, London: Macmillan.

Rossiter, M. (1974), 'Women scientists in America before 1920', *American Scientist*, 62, 312–23.

Rubin, G. (1975), 'The traffic in women: notes on the "political economy" of sex', in *Toward an Anthropology of Women*, ed. R. R. Reiter, London: Monthly Review Press.

Slater, P. J. B. (1977), 'Sociology and ethics', *Bull. of the British Psychological Society*, 30, 349–51.

Swann Report (1968), *The Flow into Employment of Scientists, Engineers and Technologists*, Cmnd 3760, London, HMSO.

Torrance, E. P. (1963), 'Changing reactions of pre-adolescent girls to tasks requiring creative scientific thinking', *J. of Genetic Psychol.*, 102, 217–23.

White, M. S. (1970), 'Psychological and social barriers to women in science', *Science*, 170, 413–16.

Women and Science Collective (1975), Special issue of *Science for People*, 29.

Zuckerman Report (1961), *The Management and Control of Research and Development*, London: HMSO.

RESEARCH STUDIES

7

Factors differentially affecting the science subject preferences, choices and attitudes of girls and boys

M. B. ORMEROD

Over the past few years a group of research students and I have conducted a series of investigations into the subject preferences, choices and attitudes of schoolchildren.[1] None of the researches set out to investigate sex differences as a primary objective. It is, however, axiomatic in any study to break down the sample into reasonable sub-samples and scrutinise the differences between these sub-samples on all the variables being measured. Thus it was entirely logical to record differences between the sexes and to examine the differences between the single-sex and co-educated pupils. Some of these differences turned out to be quite marked, and the most important of them are summarised in this chapter. Most of the investigations have already been reported in academic journals, and the reader is referred to these for further details.

Whether or not a pupil in the most able quartile of the population can ultimately take up a career based on the sciences is determined by subject choices made at fourteen years of age—i.e. at the end of the third year of secondary education—and in the case of the most able often a year earlier. In no other Western democracy is the die cast so soon. Accordingly, most of the research reported here has concentrated on able children in the thirteen- to fourteen-year-old age group, who are approaching this crucial decision. However, from scattered hints in earlier work (reported in Ormerod with Duckworth, 1975, section E) it was realised that some of the factors influencing science choices operate at even earlier ages. The last two studies discussed in this chapter trace some of the sex differences which are evident in primary and middle school science.

The effects of single-sex and co-education on science preference and choice

Previous research, summarised by Dale (1974), has indicated that

Table 7.1 The gender spectrum of common school subjects

	Chemistry	Physics	Maths	Geography	Music	Biology	History	Second foreign language	Latin	Art	French	English	Religious instruction
C.R.	9·3	8·2	4·5	0·6	0·4	0·5	1·0	1·1	1·5	2·5	6·8	7·1	7·3
% boys	70	79	61	55	36	37	48	41	50	43	46	43	37

Male ⟵ ↑ Neutral point ⟶ Female

Notes

1 C.R.= critical ratio yielded by Mann–Whitney U test for difference between sexes in subject preference. The critical ratio is the sex difference for each subject standardised by dividing by the standard error of the difference.

2 % boys = percentage of 1972 O level GCE entries from boys.

girls in single-sex schools are more likely to choose physics or physical science and mathematics than their co-educated sisters. In order to examine this phenomenon more closely I used a nation-wide sample of 1,200 fourteen-year-olds in the top 25 per cent of the ability range which included a balance of co-educational and single-sex schools. At an early stage in the research I compared the ranked preferences and choices of pupils in co-educational and single-sex schools for all the school subjects studied at the fourteen-plus stage (Ormerod, 1975). Most of the results strongly supported the suggestion that there is a polarisation of preferences and choice among the co-educated pupils when compared with those in single-sex schools. This manifests itself as a decrease in preference and choice of subjects of the opposite 'gender' to that of the pupils concerned and an increase in preferences and choice of subjects of the same 'gender' as the pupils concerned. The only exception to this pattern was girls' subject choice (discussed below).

The gender of the school subjects was determined by the difference between the sexes in preference for that subject; it is shown in Table 7.1. Thus for the 'male' sciences (chemistry or physics) and for mathematics it was found that co-educated boys showed *greater* preference than single-sex-educated boys, and co-educated girls showed a *lower* preference than single-sex educated girls. For biology, which is slightly 'female' at this age, the exact opposite was the case. For all these subjects the preferences of co-educated boys and girls were further apart than those of single-sex-educated boys and girls.

These findings gained strong support from sixth form choices reported by the Department of Education and Science (1975) on a huge sample of 447 schools. These latter results may be a consequence of earlier choices at fourteen.

Elsewhere in this book, and later in this chapter, evidence is adduced that several factors are operating to deter girls specifically from the study of physical science. But the widespread polarisation of preference in co-educational schools across almost all subjects in which these special factors are *not* operating is consistent with one all-pervading hypothesis. This is that each sex, when educated with the other, is at puberty almost driven by developmental changes to use subject preference and, where possible, subject choice as a means of asserting its sex role.

Whatever the reasons, the current trend towards more co-education will probably lead to fewer girls and more boys choosing physical sciences and mathematics and more girls and fewer boys choosing biology.

The results did not, however, support the polarisation hypothesis for girls' subject choice. In the co-educational situation there was a significantly lower correlation between preference and choice across all subjects for girls than there was for boys.[2] Thus boys' choices were more likely to match their preferences than were girls', which suggests that the options system was more suited to boys' pattern of preferences than to girls.[3] This anomaly drove Keys and Ormerod (1976a) to look more closely at the relationships between the science preferences and choices of fourteen-year-olds. With two different samples of potential O level pupils, we found a significant proportion of girls who 'liked' physics (and in one sample chemistry as well) who were dropping it, together with a significant proportion of girls who 'disliked' biology and yet were taking it.

The reasons for this situation are undetermined but could be a combination of:

1. Girls' insecurity at entering what appears to be a male preserve, coupled with
2. Their anxiety about the physical sciences' apparent difficulty (see below).

3. Advice against such choices from Heads, staff ('You'll get better grades in [e.g.] history', etc.), parents and peer groups.
4. Hidden 'selection' because the school has only enough science staff or laboratories for one physics or chemistry set. This may also motivate the supposedly dispassionate advice in (3).

The effects of perceived difficulty on girls' physical science preferences and choices

There is abundant evidence documented elsewhere (Ormerod with Duckworth, 1975) that physics and chemistry are more difficult than other school subjects both in reality and in the eyes of pupils of both sexes. Keys and Ormerod (1976b), again working with fourteen-year-old pupils in O level streams, found a strong positive correlation between girls' ranking of school subjects in order of preference and their ranking in order of perceived easiness, whilst the corresponding correlation for boys was low and not significant. This suggests that, at this critical age of subject choice, girls' subject preferences are much more strongly related to their perceptions of subject difficulty than are those of boys. This is of particular importance in the case of the notorious relative difficulty of physics and chemistry.

In a further investigation with the same sample, Keys and Ormerod (1977) obtained correlations between subject preference and standardised measures of attainment in the pupils' most recent school examinations. Across all subjects the correlations between subject preference and attainment were less than those for subject preference and perceived difficulty. The correlations between subject preferences and attainment revealed no significant sex differences. We concluded that strategies of making examinations 'easy' on subject matter which is perceived as 'difficult' are unlikely to appease anxieties about difficulty, which, as has been shown, seem to operate more strongly in the case of girls. We also suggested that the difficulty of physics and chemistry in the eyes of girls goes a long way to explain why these two subjects are found at the extreme 'male' end of the gender spectrum.

The influence of attitudes to the social implications of science on science choices at fourteen

A country-wide sample of 2,200 fourteen-year-old pupils who were potential O level candidates was used to study the influence of attitudes on subject choice. Ranked subject preferences, subject choices for the following year, and measures of teacher liking were obtained from this sample. In addition, an attitude instrument was used. This yielded five independent scores as follows:

1. A 'SUBATT' score measuring attitudes to science or the sciences as school subjects;

and four 'SOCATT' scores concerned with attitudes to the social implications of science:

2. An 'AESTHETIC/HUMANITARIAN' score concerned mainly with the real or supposed effects of science on the environment or the human race.
3. A 'PRACTICAL' score concerned with the practical value of science to the individual.
4. A 'MONEY' score concerned with the value of science to the community, usually measured by reactions towards money spent on science by the state.
5. A 'SCIENTIST' score—concerned with pupils' feelings about the characters and activities of scientists.

Details of these scales and their interrelationships are reported elsewhere (Ormerod, 1979). The most important findings concern the relationships between the SOCATT scores and science choices.

First, there are relatively few significant correlations between biology choice and views on the social implications of science, whereas low but significant correlations with physics and chemistry choice abound. In other words, pupils in general have mildly favourable attitudes to the social implications of science if they choose to study physics and chemistry beyond fourteen-plus, but these are not necessarily present in the case of biology. Put another way, unfavourable attitudes to the social implications of science seem to divert pupils from the further study of physics and chemistry but not from biology.

Second, for both sexes there are low but significant correlations between the outcome measures (science preference and choice) and favourable attitudes on the MONEY and SCIENTIST scales. In the other two SOCATT scales, however, there is a notable sex difference. With boys a favourable outlook on the PRACTICAL pay-off of science to the individual correlates significantly with physics and chemistry choice, but with girls it does not. With boys, favourable views on the AESTHETIC/HUMANITARIAN aspects of science are unconnected with physics and chemistry choices. With girls this latter cluster of views is clearly associated with choice of the physical sciences.

Lastly, the relationships of attidude scores to science subject choices are modified by the effects of teacher liking. These relationships are complicated, but over the whole range of physical science, it seems that boys' attitudes to science only affect their subject choice when they like the teacher. For girls the exact opposite often occurs. When they like the teacher a favourable attitude has no effect on girls' choice of science, but a favourable attitude to the subject can compensate for dislike of the teacher.

Thus it appears important to deal with a range of aspects of the social implications of science for both sexes before the critical phase of science choices at fourteen. Attempts to improve aesthetic/humanitarian aspects of attitudes to science are likely to be of more importance for girls and 'practical value to the individual' for boys. Even here, however, favourable attitudes can offset dislike of the teacher by girls.

'Social implications' are creeping into science syllabuses but often are only mentioned after choices have been made.

Middle school science and its effects on later biology, chemistry and physics preferences and choices

Jennifer Bottomley has monitored the effect of middle school science for 623 pupils in fifty-three middle schools for pupils of eight to twelve who became the intake of four secondary schools—a boys', a girls' and a co-educational grammar school and a secondary modern school. The first phase of the study centred on a list of forty-one science activities which the pupils recollected having done or not done in their middle schools, together with an indication of their liking for middle school science and their teacher(s) of it (Bottomley and Ormerod, 1977). The results indicated that there was a variety of activities whose recollected performance or omission correlated over a range of +0·3 to –0·3 with expressed preference for middle school science. Superficially there was no apparent sex difference between these activities. In particular, biological activities did not emerge as being more strongly correlated with middle school science liking among girls, as might have been expected. A more detailed examination of the specific biological activities, however, revealed that the six with the highest correlations with science liking among the girls were all botanical; collectively they distinguished the girls from the boys significantly.

A more thorough analysis of the relationships between whether pupils have or have not experienced a range of biological activities in middle school and liking for biology after one year of secondary school reveals that having pursued certain *specific* activities concerned with live animals (keeping snails and woodlice in the classroom, studying the growth, movement and life cycles of any animals, drawing animals from life) correlates *negatively* with biology liking *among those girls who have decided whether or not to continue with biology*, i.e. once those who are uncertain about subject choice have been eliminated. These correlations differ significantly from the positive or zero correlations found for boys. In sharp contrast, the correlations of the two activities in which pupils have a nurturative role—feeding and cleaning animals—are *positive* for girls and *negative* for boys. Thus already at some age between about nine and twelve the scope for exercising scientific curiosity about animals is attracting boys to biology whilst the physical contact that this involves is offputting to girls. At this age girls are already adopting traditional feminine reactions towards proximity with animals other than domestic animals and also a 'feminine' nurturative role in feeding and cleaning.[4]

These pupils were also given a questionnaire after a year at secondary school, i.e. at the age of thirteen. The questionnaire consisted of a range of items such as liking for the separate sciences, mathematics liking, liking for practical work in each science; liking for the teacher of each science; perceptions of the difficulty of each science; mothers' and fathers' perceived interest in science. In addition they were asked, 'If you could choose to continue to take or could drop each science now, what would you do—take, uncertain or drop?' There was an 'exam guess' as to whether the pupils thought they would do badly, average or well in the coming school examinations, and a father's and mother's science choice score indicating whether the pupils thought their parents wanted them to continue to study each science subject.

Discriminant function analysis was used with these data to identify the variables which were most strongly related to subject liking and choice.[5] Some variables were revealed as influencing the preference or choice of one sex only, others had a stronger influence for one sex than the other. If anything, the choices showed more of a pattern of differences than the 'likings', but in a rough-and-ready sort of way the sex differences can be summarised as below. In considering these results one must bear in mind the 'feminity' of biology at this stage and the 'masculinity' of physics and chemistry.

The order of science subject liking was:

Boys: chemistry > physics > biology
Girls: biology > chemistry > physics

With this in mind we can examine the sex differences under several headings, as follows:

1. Difficulty and subject liking

In no case does difficulty appear as a significant predictor of subject liking for the most preferred subject (chemistry for boys and biology for girls). Difficulty is the lowest of the significant predictors for girls' chemistry. It appears for both boys and girls as significant in determining physics liking but is a much more powerful discriminator for girls' physics liking than it is for boys'. Also it is much more powerful for girls' physics liking than for girls' chemistry liking.

Another factor is entering the subject here, namely the recognised inherent difficulty of physics associated with the introduction of quantitative concepts from an early stage whilst chemistry is still descriptive. Thus the 'male' subject mathematics has become involved. Mathematics liking turns up as one of the weaker positive discriminators in girls' biology choice. To put the latter point more succinctly, girls who like mathematics are more likely to choose physics and drop biology.

Apart from 'intrinsic' difficulty, the general situation seems explicable in terms of either

(a) Pupils' unwillingness to take difficulties in their stride when associated with subjects of opposite 'gender'.
(b) The possibility that in the co-educational situation biology is being taught in a style which makes it less difficult for girls and the converse is the case for physics and chemistry. This would fit in with Galton's results (see Chapter 13).

One of the curses of science education in co-educational schools *so long as early choices are offered* is the strong temptation for the physics or chemistry teacher to assume by year three or even earlier that 'Most of my class are going to be boys by year four'. Hence there is an unconscious incentive for the physical science teacher to concentrate on boys and teach in a style more suited to boys *even before choices are offered*, thus aggravating a pernicious cycle. The reverse may be true of biology teaching, thus reinforcing girls' predilections, and could well be occuring right across the 'gender spectrum' of optional subjects.

2. Practical work

There is also an important sex difference in the position of practical work. In all three subject choices 'liking of practical work' appears as a significant discriminator for boys but not for girls. This does not imply that girls dislike practical work or are put off by it. In the analysis of subject liking it appears as one of the strongest discriminators for both sexes. In the choice situations, however, girls' liking for practical work is completely accounted for 'within' their liking for the subject. Boys on the other hand have in practical work an extra incentive for choosing all three sciences which the girls have not.

There is a good deal of supportive evidence from American investigators that practical work is less attractive to girls and presents them with greater learning problems. Thus it has been found that girls of seven learn better from watching demonstrations than by active involvement (Fogelman, 1969). At eleven girls are more dependent on cues from the teacher (Ogunyemi, 1972). Yet Rains (1970) found that in practical classes of pupils of the same age there are significantly more interactions between boys and the teacher. We have already noted how middle school science activities involving animals are counter-productive for the future liking for secondary school biology among girls. These facts should be considered in the light of the recent decade of science curriculum reform which has placed a lot more emphasis on practical work. Is this not also indicative of an unconscious prejudice in favour of teaching to interest boys?

3. General effects of middle school science on the secondary sciences

'Liking for middle school science' (as recorded at the beginning of secondary schools) only persists as a significant predictor for secondary school science in the case of girls. It is a *negative* predictor for biology liking, a positive predictor for chemistry choice and *negative* predictor for physics liking. This would imply that the middle school science effects persist for girls and are pro-chemistry and anti-physics. As far as physics is concerned this seems to justify the fears of the ASE (1976) document on the staffing of middle schools. These were that the science education therein would be mainly biologically oriented, with little physical science. The evidence also points to the early establishment of a biological bias in girls' science education—a point which also comes out strongly in

Charles Wood's study (see below).

The presence of middle school science liking as a *negative* predictor of secondary school biology liking among girls would be surprising had we not found how off-putting work with animals could be to the fair sex.

4. Parental interest

Parental interests in science and parental science choice scores were frequently significant discriminators of science liking and choice. This emphasises the importance of parental influence, but there is some indication that they are operating to direct the sexes in traditional directions, sometimes with opposite results to those intended. 'Father's interest in science' acts as a significant 'negative' discriminator in boys' chemistry and physics liking, and mothers had better stick to knitting if they want their sons to choose physics. This is not all as nonsensical as it sounds. In my twenty-two years of teaching I have seen quite a few cases of parents causing considerable unhappiness by transferring their own ambitions to children. Pupils can be led but not driven, and it is easy to kindle resentment rather than interest if parental ambition is out of line with the pupil's own interest and abilities.

Science attitudes in the primary school

With Charles Wood, a primary school headmaster, I have recently constructed a science attitude scale for use with children aged nine to eleven. This consists of statements drawn from the free expression of pupils of this age range on areas of science designated by experienced primary school teachers as likely to be of interest at this age.

A group of fifty-five statements were selected in this way and administered to 290 pupils. A range of five responses from 'strongly agree' through 'uncertain' to 'strongly disagree' was used. Factor analysis revealed two strong clusters of items, and the statements were grouped accordingly to form a 'space' scale and a 'nature study' scale.

The 'space' scale was mainly concerned with astronomy and interest in space exploration. The high loading of 'interest in dinosaurs' on this scale rather than with the 'nature study' scale strongly suggests that this sort of interest has been kindled by TV and film viewing of 'space sagas'. Not surprisingly, boys scored very significantly higher than girls on this scale, whereas on the 'nature study' scale the scores of girls were significantly higher than those of boys. This result is strongly indicative of the early sex-related bifurcation of science interest into physical and biological fields.

The result cannot, however, be claimed to support either the 'nature' (Chapter 3) or the 'nurture' (Chapter 4) hypothesis except in highlighting the fact that nurture must be operating very early in children's lives.

Conclusion

In this chapter I have tried to demonstrate the complexity of the influences on girls' and boys' subject choices and preferences. Many of these influences are subtle and unstable, clearly manifest in some situations but absent in others. Future researchers must therefore be careful to separate out the different situations, such as liking or not liking the teacher, being in a single-sex or a co-educational school, studying or caring for live animals. It is also important to distinguish the different branches of science, particularly biology and physical science, since pupils perceive these subject quite differently.

This study has confirmed previous work showing that girls have more favourable attitudes to physical science in single-sex schools than in co-educational schools. It has also shown that girls dislike subjects which they perceive as difficult, and this applies particularly to physics. Views on the social implications of science are largely irrelevant to pupil choice of biology, but important in choosing physical science; the aesthetic/humanitarian aspects of science are particularly important for girls, but boys are more influenced by their views on the practical value of science to the individual. Boys and girls in the nine to eleven age range have clearly differentiated interests in science, with girls preferring nature studies and boys preferring 'space' sciences. In the same age range, undertaking botanical activities in middle school was more strongly related to liking science for girls than for boys; however, girls who had done experimental work with animals in middle school tended to dislike biology at a later stage.

In the light of these research results, what steps might be taken to induce more girls to specialise in science? One possible reform would be to postpone subject choices. Fourteen is a most undesirable age for making subject choices, since non-academic reasons such as gender identity frequently predominate. Most reorganisation of schools involves more co-education, which only accentuates the problem, and suggest that in the future even fewer girls than at present will opt to study physical science. It is entirely reasonable to argue that both sexes ought to have a balanced science education with both a physical and a biological content up to the age of sixteen.

However, to make physical science compulsory for girls to sixteen would be disastrous without revision of syllabuses in two directions:

1. The level of difficulty must be reduced to bring physical science into line with other subjects.
2. In place of the difficult material removed or postponed there will have to be serious efforts to emphasise the importance of science to our society, *and by the age of fourteen*, to improve its tarnished image as a menace to human well-being and natural beauty. This is not going to be easy. To some extent it requires a knowledge of social history to paint a picture of the miseries of disease and the intellectual poverty suffered by all but the most favoured of past generations. It would be a great help if the industries heavily dependent on science could be persuaded to make films and other aids to learning to help in this area.

Girls are being taught physics and chemistry successfully in some schools. In other schools teachers (unconsciously or otherwise) could be teaching the boys rather than girls. Research needs to be undertaken to see whether such styles are discernable. If this is the case, compulsory physical science up to sixteen will never prosper without the dissemination of experience and research in the form of short local courses. Such courses could with profit also consider ways of dealing with the social implications of science.

Notes

1 I would like to express my gratitude to the research students, Wendy Keys, Jennifer Bottomley and Charles Wood, whose results are incorporated in this chapter. Without their industry and initiative no such comprehensive findings would have been possible. The repeated refusal of external funds to support this work has meant that its continuation has been dependent upon the dedication of research workers subsisting on SSRC studentships or doing unpaid part-time work.

2 See the Statistical Appendix for a discussion of correlation and statistical significance.

3 A similar implication can be drawn from the reasons given by girls and boys in Chapter 9 for dropping or continuing science subjects. [*Editor*]

4 *Cf.* the pupils' discussion on dissection in Chapter 15. [*Editor*]

5 See the Statistical Appendix for a brief discussion of discriminant analysis.

References

Association for Science Education (1976), *The Supply of Science Teachers*, Hatfield: Association for Science Education.

Bottomley, Jennifer M., and Ormerod, M. B. (1977), 'Middle school science activities and their association with liking for science', *Education in Science*, No. 74, 23.

Dale, R. R. (1974), *Mixed or Single Sex Schools*, London: Routledge and Kegan Paul, Vol. 3.

Department of Education and Science (1975), *Curricular Differences for Boys and Girls*, Education Survey No. 21 London: HMSO.

Fogelman, K. R. (1969), 'Piagetian tests and sex differences', *Educational Research*, Vol. 12, 154–5.

Keys, Wendy, and Ormerod, M. B. (1976a), 'A comparison of the pattern of science subject choices for boys and girls in the light of pupils' own expressed subject preferences', *School Science Review*, Vol. 63, No. 203, 348–50.

—(1976b), 'Some factors affecting pupils' subject preferences', *Durham Research Review*, Vol. 36, 1109–15.

—(1977), 'Some sex-related differences in the correlates of subject preference in the middle years of secondary education', *Educational Studies*, Vol. 3, 2, 111–16.

Ogunyemi, E. L. (1972), 'The effect of different sources of verbalized information on performance at a science-related cognitive task', *J. Res. Sci. Teach.*, Vol. 9, 157–65.

Ormerod, M. B. (1975), 'Subject preference and choice in co-educational and single sex secondary schools', *Brit. J. Educ. Psychol.*, Vol. 45, 257–67.

—(1979), 'Pupils' attitudes to the social implications of science', *European J. Science Education*, Vol. 1, 177.

Ormerod, M. B., with Duckworth, D. (1975), *Pupils' Attitudes to Science: a Review of Research*, Slough: NFER. Section D.

Rains, O. W. (1970), 'A study of teacher–pupil and pupil–pupil interactional differences between inquiry-centred science and traditional science in elementary schools', unpublished Ed.D. dissertation, Oklahoma State University.

8

Science options in a girls' grammar school

DAVE EBBUTT

During the spring of 1977 I carried out a case study in science choice at the request of the headmistress and science staff of a four-form-entry girls' grammar school. My remit was 'to investigate the factors which influence third year girls' choice of science subjects for their fourth and fifth year options'.

My task as I saw it was twofold. Firstly to familiarise myself with the options system and to find out in terms of science subjects who was doing what—the numbers game. The second task—in a sense, what lies beyond the numbers game—was to attempt to find out why. It is this area upon which I shall focus for the main part of this chapter, after the briefest consideration of the option system.

Taking the tasks in turn, the first step was to represent the curriculum in a form I could handle. I chose the diagrammatic form adopted by Reid *et al.* (1974), and Fig. 8.1 shows the curriculum structure in the study school. In the first two years all the girls are taught combined science, and in the third year they are all taught biology, chemistry and physics as separate subjects. After the third year they encounter an option system which is similar in form to that in the majority of secondary schools. They are asked to make a preliminary choice of one subject from six in each column. This selection is subject to a number of constraints; for instance, one science (physics, chemistry, biology or physical science) is compulsory. The choice involves selecting four subjects from thirteen together with a fifth as a 'spare'. The girls then enter their selections on another sheet which is collected by third form teachers. The choices are tabulated form by form, and if viable teaching groups emerge, then the options are said to be 'first choices'. If viable groups do not immediately emerge, then presumably one or more of the anomalous selections is renegotiated until the options are finalised.

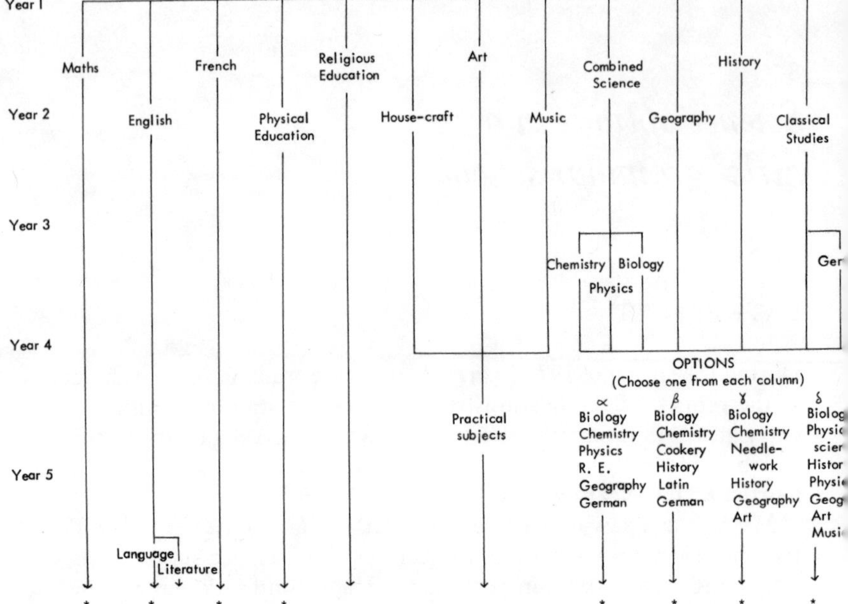

Fig. 8.1 Study School: curriculum (years 1 to 5). *Public examinations.

When I had a clear picture of the system I obtained information from the fourth-year tutor as to how the current (i.e. 1977) fourth year had opted within these alternatives. Subsequently I was able to compare these data with those coming in from third-year girls who were in the process of making their choices for the future. I was then in a position to compare the pattern of option selection between two year cohorts and to see how this pattern compared with the national average for girls in girls' grammar schools.

Table 8.1 shows that the girls in the study school were opting for the various sciences in a rank order of popularity which is replicated nationally. Further, the girls in the study school were opting for science subjects in a numerical pattern which closely corresponds to the national picture for similar types of school.

Having completed my first task to my satisfaction, I rephrased the second task in my mind as 'What factors are operating in the study school which might also be generalisable nationally which give rise to the observable patterns of choice?'

I decided on a three-pronged approach. I began with taped interviews with each of the seven science staff (two men, five women)—having previously negotiated a confidentiality

Table 8.1 The percentage of girls taking science subjects in the study school, compared to the national pattern for girls' grammar schools.

Subject	*National pattern* (*% taking in years 4 and 5*)	*Study school* (*% taking in year 4*)
Physics	33	34
Chemistry	40	47
Biology	70	77

Source: Department of Education and Science, *Curricular Differences for Boys and Girls*, HMSO, 1975.

agreement—to elicit their views on the factors behind the patterns of choice. It was impossible to transcribe all this material, but on each subsequent visit to the school I was able to present a set of 'edited highlights' of each of the conversations with the teachers concerned for their approval, alterations or rejection. At the conclusion of this phase I had seven sets of 'theories' as to why the science staff thought the girls had opted as they had. The binding nature of my confidentiality agreement does not allow me to include the seven 'theories' here, but analysis of these documents 'revealed' the following trends of thought.

There appeared to be complete agreement among the science staff that biology is easier for girls than the other sciences. Moreover five of the teachers went a stage further and identified criteria upon which this perceived easiness was based. They suggested that it was the 'predictability' or 'straightforwardness' of biology which makes it easier. An alternative viewpoint was that it is the quantitative nature or the mathematical content of the other sciences which is seen as difficult by the girls (three teachers). One teacher extended this further to embrace the distinction between learning and understanding: he held that the study of the physical sciences necessitates a greater understanding and interpretation, whereas biology is a science which can be learned and memorised, and commented, 'I can see therefore why it is popular in a girls' school.' Two of the staff adopted this idea and applied it to the option system. They suggested that the girls may be making a positive choice for physics because they perceive it to be difficult and, in some cases at least, a negative choice for biology because it is perceived to be easier.[1]

Another factor noted by several of the staff was that biology is intrinsically more interesting to girls, and these teachers tentatively identified the girls' focus of interest as 'human anatomy' or 'a relevant and functional guide to themselves'.[2]

All the science department perceived as important the career implications behind subject choice. Related to the pupils' career orientation, and again universally held by members of the department, was the view that the girls who come to the school are almost predisposed to chose biology, because they are who they are, the daughters of parents who expect the school to teach biology (and the school strives to meet these parental expectations). Thus parental influence was seen by the science staff, whether consciously or unconsciously perceived by the girls themselves, to be an unknown but important factor in girls' subject choice.

An interesting view was held by one of the biology teachers who suggested that biology as it is taught in the school matches, suits or is in tune with the developmental stage of the girls. This was seen in terms of the girls' 'academic ambition, bookishness, being a swot and doing well'. Other minority held views by members of the department as to other possible factors included the influence of friends, the Junior school from which the pupil had originally transferred, and the passion which some girls expend on horses or ponies—hippomania.

In talking to all the teachers in the science department I was able to put some flesh on the bones of yet a further factor which three of the teachers explicitly saw as influencing the girls' subject choice (it was alluded to by others)—the influence of the teacher and the teaching style, particularly in the third year, which I saw reflected in the views that the various subject teachers held as to the content, aims and objectives of the third-year courses. The physicists, for example, did not see the third-year course as either 'part of the strict O level course' or as a 'good predictor' of what physics would be like later. Physics is taught in the third year from an experimental, problem-solving point of view, but at option time the teacher emphasises mathematical skills to those seeking to opt for the subject.

On the other hand the basis of third-year chemistry was stated as a foretaste of 'the sort of approach they're going to run into', and the children are exposed to the 'complexity' of the subject. Whereas biology—virtually a new subject in the third year, since the biological component of combined science in first and second year is neglible—was seen by the biology teachers as explicitly part of the O level course. But the easier, 'more interesting' topics—human biology—are covered in the first term and a half of year three, whilst the 'more abstract and difficult processes'—osmosis—are not covered until *after* the girls have made their subject choice. Moreover the syllabus followed and the teaching techniques employed conceive of biology as non-problematic.

Armed with these theories and insights I decided to investigate the degree to which the ideas expressed by the science staff, as to the factors they thought might be affecting the girls' choice of science subjects, were corroborated or otherwise by the girls themselves. During the lunch breaks I interviewed some fourth-year girls from two forms in their form rooms. Each time I entered the room I sat down away from the main chat centres and waited (they had been told I was around). Sometimes girls came forward spontaneously; in other instances, and more usually, I asked the girls with whom I was talking to invite somebody over who was taking a specific combination of subjects. Always I asked if I might use the cassette recorder and promised to wipe the tape clean when I had finished the documentation. I also promised anonymity. Each speaker was identified in the analysis by her subject combination only.

The majority of the girls I spoke to conceived of biology as the easiest of the sciences (I did not ask all the girls the same questions), more or less confirming the feelings held by the teachers. All the pupils who I asked to rank the three sciences in order of difficulty gave physics as most difficult, followed by chemistry, with biology as the easiest. The majority also ascribed the source of the difficulty of physics to its mathematical or quantitative basis. 'Most people think biology is the easiest and physics the hardest because of the maths involved' (taking three sciences). 'Didn't do physics 'cause no good at maths' (taking biology only). 'Considered physics, but too much maths' (taking biology only). 'I hadn't got on well . . . not mathematical' (taking biology and chemistry). One person saw this as the attraction of physics: 'I prefer to work with numbers and symbols rather than words' (taking physics only).

There are echoes of the teachers' distinction between learning and understanding in the following comments: '. . . facts in physics have to be used' (taking physics only). 'In chemistry you have to know what you're doing' (taking biology and chemistry). 'Physics and chemistry are cumulative' (taking three sciences). These two notions were woven together by a girl who intended to be a doctor when she explained her rank order of difficulty: 'It's the calculations and understanding and using the ideas of physics'. Three girls agreed that the predictability of biology was a factor in choice, but the word 'predictability' was always used first by me: nobody volunteered it as a factor spontaneously. I cannot decide whether I generated the idea or merely supplied a label for a notion they had themselves.

The evidence furnished by the pupils with regard to the teachers' hypothesis that physics is a positive choice was equivocal. Only one of the three girls taking physics only with whom I talked made her choice on unambiguously positive criteria: 'My dad and I thought of it together; I took physics as a challenge . . . it didn't close career options'. The second girl stated: 'Didn't like chemistry and biology, didn't understand them,' but she was the girl who preferred to work with numbers and symbols above. The third girl had definitely made a negative choice: 'I enjoyed physics a lot, it was my least worst, I would have dropped sciences completely if I could.' All the other physicists were taking three sciences, and all had chosen medical careers; thus their choice of physics can be said to be positive.

I was able to locate only two other clear-cut negative choices, one from a pupil taking biology: 'Didn't want to do any, really. Too much like hard work, really. I've felt like this ever since my first science lesson'. The other case was a girl taking biology and chemistry: 'I wanted to do A level biology, so had to do chemistry.' As a factor involved in the positive choice of biology, I gathered some evidence to suggest that at least some girls were seeking, to use a phrase from the teachers' analysis, 'a functional guide to themselves'. 'Biology is learning about yourself—more realistic—answering questions about yourself' (taking biology only). '. . . explains you to yourself' (taking biology only). 'Biology answers questions about you' (taking biology only). Alternatively there were examples where biology was thought to be intrinsically more interesting or that interest was specified '. . . more interesting' (taking three sciences). 'Biology is easier and more interesting' (taking biology and chemistry). 'Girls are more interested in reproduction' (taking three sciences).[3]

Career possibilities were seen as a vital factor by all the girls, and if a career had not been decided upon, then keeping career options open was seen to be important. Cuddly animals, hippomania, 'the animals thing' were totally and universally rejected as a factor in choosing subjects, the pupils pointing out that this 'stage' is over by the final year of junior school or certainly by the end of the first year of secondary school. Similarly none of the girls considered her friends to be an influence in her choice of subject, and many made the point that, whilst knowing which subject the friend had opted for, there was no guaranteed way of knowing which teaching group either had been designated to. A similar constraint applied in the teacher-liking dimension; whilst admitting that 'Of course there are teachers you like' (taking three sciences). it was only in the case of physical science that 'teacher liking' became a real factor in that there was only one group in year four for that subject. 'I was not influenced by friends. I liked the teacher' (taking physical science only). This is not to say that the pupils were not aware of differences in teaching style and content on a subject basis in year three. It was widely noticed that the physics course, and to a lesser extent the chemistry course, in year three bore little comparison with those in year four. 'Biology is sitting and being told' (taking biology only). 'She dictates so we don't get anything wrong' (taking biology only). 'Biology in the third year is easier because of the way it's taught, this accounts for the imbalance' (taking three sciences). 'Chemistry in year three is not at all like chemistry in the fourth year—there is no practical in the fourth year' (taking three sciences). 'Physics in the third year is no guide to physics in the fourth year' (taking physics only). 'Third-year physics doesn't really relate' (taking physics only).

Many of the girls hinted that there was an element of parental guidance in the choice process, ranging from 'They didn't really mind' (taking biology only) to the girl already mentioned who decided with her dad to take physics as a challenge. Another girl stated: 'My parents thought I would be OK for a job' (taking three sciences). Finally one or two extra opinions which I found interesting: 'Chemistry is the most impersonal' (taking three sciences). 'Female scientists tend to be frumpy, but nurses tend to be glamorous' (taking three sciences). 'I haven't an image of a scientist . . . not particularly glamorous; nurses are' (taking three sciences). It was apparent from this analysis that there was considerable overlap or congruence between the teachers' and the pupils' responses. Only with respect to two 'factors', *viz.* friends and 'the animals bit', did the girls' comments negate the equivalent teacher hypothesis.

The third prong of my approach was an attempt to 'test' the third-year girls' perception of the roles women scientists might fulfil in a given situation. At the same time I was attempting to uncover stereotypes which the pupils might have had in their minds with regard to the women who occupy those roles. The Third Year Tutor in the English Department kindly organised the setting of the following essay during normal scheduled English time. The girls were unaware that it was for me or would form part of my study.

> *Time.* The near future.
> *Place.* Somewhere in the Third World.
> *Situation.* A huge natural disaster: the UN sends in scientific teams to assess the situation. Two of the teams are led by women.
> *Task.* Describe the women and their jobs.

The essays were a 'good read', involving drought, earthquakes, tidal waves and pestilence on a scale that would have made Cecil B. De Mille blanche. Turning to the description of the scientists (which I am aware were to an extent prespecified by the fictional situation and locale), the Third Year Tutor noted, as a result of a glance through their work, that 'The fact that this was done in an English lesson probably suggested some attempt at "fine writing" '. She went on, 'lots of thick glasses, flat shoes, big feet, judo types with muscular calves and sensible clothes'. I have selected several examples of this genre:

> Mrs Smythe was a burly woman with a loud voice and she wore large dark rimmed glasses that swamped her face. She was only 5ft. 1in. tall and so everyone looked down at her. She had short curly hair and thick eyebrows. Every other word she said was connected to science; she had to give technical details and chemical equations which made her a bore.

> Miss Harper was an immaculate person. Her hair was tied back neatly into a bun and her clothes showed no signs of a crease. Everything about her was orderly, everything scientific she did was recorded. All her belongings were labelled neatly.

> Mrs McDougall was a tall very slim limbed person, she had sharp facial features which gave her a rather strict stern appearance. She had bright green beady eyes which never missed even the slightest thing and long red hair which was always tied back in a bun. Her qualifications, the fact that she was sensible and always kept calm cool and collected, this plus the fact that she had the endurance of the toughest man, made her very recommendable.

The Third Year Tutor in her brief *résumé* after a swift scan of the girls' work concentrated on the physical appearance of the heroines. Admittedly these do seem to be characterised by qualities other than conventional beauty. What, apart from the media, could be the source of these physical stereotypes? Certainly not the Head of Science of the school (a physicist), for she radiated elegance and poise. Perhaps this concentration on physical appearance was wrongly placed, because other qualities are also emphasised in the passages quoted and in the rest of the written work. What came through most strongly to me was the descriptions of how sensible, capable, reliable, cool, calm, logical, methodical, positive and orderly, exact, technical and dedicated were the women portrayed. These attributes I suggest, are not the characteristics which the average fourteen-year-old post-adolescent girl would espouse as part of her own self-image. Indeed, drawing upon my experience of teaching this cohort, I would speculate that the values mentioned are almost diametrically opposite to the self-image of most third-year girls.

I would wish to rest my case by speculating that the fictional stereotypes revealed in the girls' written work may reflect their perceptions of the 'real' world. If this were so, then it could well reflect back into the option system, which is theoretically 'open' to any girl with the necessary skills. But the option system may not be quite as open as it seems. As I have shown, career implications were an important factor influencing the girls' choice of science subjects. If women in science-based careers are seen to require such characteristics as capability, logicality and exactitude, there may be considerable self-selection away from science subjects by girls who do not see themselves as possessing these traits at fourteen years of age.[4]

It is beyond the scope of a qualitative study of this type to attempt to rank the various factors which influence girls' subject choice in terms of their relative importance; indeed, a study which attempted to do this would be a very different beast. Nevertheless certain of the factors are explicitly recognised to be important by the school as an institution and receive formal support in the form of mechanisms or systems. The existence of an active careers department is one such mechanism, as is the carefully planned programme of parental consultation during the option process. Organisation of the school into subject departments allows for the development of small, semi-autonomous forums in which issues concerned with teaching style, aims, content and organisation can be discussed. However, the question as to whether those factors which receive formal and systematic institutional suport are indeed the *most* important ones affecting the girls within the school still remains to be resolved.[5]

Notes

1 *Cf.* the similar implication which can be drawn from Ormerod's data (Chapter 7) and from some of the teachers' comments (Chapter 18). [*Editor*].

2 *Cf.* the pupils comments in Chapter 17. [*Editor*]
3 Both positive and negative reasons for taking science subjects are evident in the pupils' comments in Chapter 17. [*Editor*]
4 See Chapter 16 for further discussion of the stereotypes of scientists and science and the way these may affect pupils' subject choices. [*Editor*]
5 I wish to thank all members of the Science Department and particularly the Head of the school for allowing this article to be published.

Reference

Reid, M. I. Barnett, B. R., and Rosenberg, H. A., (1974) *A Matter of Choice*, NFER, Slough.

9

Choosing or channelling?

ALISON KELLY

In primary school and the first few years of secondary school, all subjects are compulsory. Pupils first encounter options at about the age of thirteen or fourteen. Previous chapters have discussed some of the attitudinal and personality correlates of science choice at this stage. Here I want to examine the patterns of choice and their consequences more closely. To what extent do choices for or against science at this age structure pupils' later educational careers? What reasons do pupils give for continuing or dropping science? Are they aware of the implications of their decisions?[1]

Data

To answer these questions I made use of the Scottish Education Data Archive.[2] Information was available from three separate but related surveys of Scottish school leavers in 1975–76. Postal questionnaires had been sent to two in five of pupils in five regions of Scotland (containing over 70 per cent of the school population) who left school in that year with no Scottish Certificates of Education (SCE) qualifications (non-certificate leavers); to two in five of all pupils in all regions who left school having presented for at least one subject at Ordinary grade but for no subject at the Higher grade (O leavers); and to two in five of all pupils who left school having presented for at least one Higher subject (H leavers).[3] More than 80 per cent of the leavers returned their questionnaires. They were asked a series of questions about their examination subjects and their school and home backgrounds. A random subsample of O and H grade leavers were asked more detailed questions about science (while other random sub-samples were asked about other school subjects).[4]

The pattern of early decisions

The idea that science should be part of every child's educational experience as well as a vocational subject for future specialists is quite widely accepted. This is particularly true in Scotland, where the value of a broad education and delayed specialisation has traditionally been recognised. But how are these ideals applied in practice? Is science in fact a part of most pupils' educational experience in their later years at school? Table 9.1 shows the percentage of pupils who studied no science (defined earlier in the question as 'physics, chemistry and biology') after the end of their second year in secondary school (the normal point at which options are chosen in Scotland). Leavers have been divided into five broad bands, according to their SCE presentations. These presentation levels may be regarded as corresponding loosely to ability levels. Nearly half of all the girls who left school without attempting any Highers (the bottom three presentation levels) had not studied science since they were fourteen. Such girls comprised nearly 70 per cent of all leavers. Even at the highest presentation level (those who attempted five or more Highers in their fifth year at school), a quarter of the girls had received no science education beyond the second year (S2). But this was true for only 7 per cent of such boys.

Table 9.1 Leavers who studied no science after second year (S2), by sex and presentation level (%)

Attempted:	*Girls*		*Boys*
5+ H grades in S5	25% (of 146)	**	7% (of 188)
1–4 H grades in S5	32% (of 385)	**	10% (of 307)
6+ 0 grades but no H grades at school	47% (of 206)	**	23% (of 246)
1–5 O grades but no H grades at school	46% (of 443)	**	25% (of 388)
No SCEs at school	43% (of 1572)	**	34% (of 1828)

** Sex difference significant beyond the 1% level. (See the Statistical Appendix for a discussion of significance level.)

There is, therefore, a very real sense in which science is a boys' subject. At every presentation or ability level a substantially greater proportion of girls than boys had dropped science before third year (S3). Possibly some of these pupils subsequently encountered scientific ideas in other subjects such as home economics or technical drawing. But the scientific content of such courses is severely limited and it is doubtful whether they could ever take the place of purpose-designed courses in science. Meanwhile a large proportion of Scottish girls (and some boys) cease the study of science completely at the age of fourteen.

The situation can be examined in more detail by considering the three main science subjects separately. This is done in Table 9.2, which shows the percentages of pupils studying each subject in the third year. Again the respondents are divided into presentation or ability levels, but non-certificate leavers are omitted because they were not asked questions about the separate science subjects. Some striking patterns emerge. It is obvious that the more 'able' the pupil the more likely s/he was to have studied physics or chemistry. But although the trend was the same for both sexes, the level of uptake of

Table 9.2 Percentages of girls and boys who studied each science subject in third year (S3) by presentation level

	Physics			*Chemistry*			*Biology*			*N*	
Attempted:	*Girls*		*Boys*	*Girls*		*Boys*	*Girls*		*Boys*	*Girls*	*Boys*
5+ H grades in S5	46	**	87	60	**	83	25		21	159	193
1–4 H grades in S5	20	**	72	41	**	66	42	**	20	422	327
6+ O grades but no H grades at school	8	**	64	28	**	52	33	**	20	184	217
1–5 O grades but no H grades at school	4	**	29	13	**	29	25		26	398	315

Notes
** Sex difference significant beyond the 1% level.

Each percentage is a percentage of the corresponding N, i.e. 46% of 159 girls who attempted 5+ H grades in S5 had studied physics in S3, and so on.

the science subjects was very different. This was particularly so for physics, which was studied by only a tiny proportion (less than 10 per cent) of girls who attempted no H grades at school but by 64 per cent of boys taking six or more O grades but no H grades. The discrepancy is almost as sharp at the highest presentation or ability level, where less than half the girls but almost 90 per cent of the boys studied physics in the third year. Rather more girls at each level studied chemistry, but they were still heavily outnumbered by boys. The pattern for biology was different. Biology was studied by about 20 per cent of boys at all levels but by a somewhat higher proportion of girls, particularly in the middle presentation levels.

I also inspected the percentages of boys and girls who studied every combination of physics, chemistry and biology in S3. Again there were marked differences between the sexes and between different presentation levels. This was most clearly apparent in the most popular (and useful) combination, that of physics and chemistry. The more able the pupil, the more likely s/he was to have studied physics and chemistry; but, at every level, boys were more likely than girls to study this combination of subjects. In general every combination involving physics was studied more widely by boys than girls, whereas the reverse was true for most combinations involving biology.

Not all pupils who studied science subjects in the third year actually attempted O grades in these subjects. Most of the H leavers who had studied each science subject in S3 went on to attempt an O grade in that subject, but this was only true of about half the O leavers. By the time they left school only 3 per cent of the girl O leavers had attempted O grade physics. Only 8 per cent of these girls had attempted O grade chemistry, and only 15 per cent had attempted O grade biology. The corresponding figures for boy O leavers were 29 per cent, 22 per cent and 10 per cent. Science was completely absent from the O grade curriculum of more than three-quarters of the girl O leavers, and the same was true for more than half the boy O leavers.

The pattern of decisions on leaving school

Despite the dearth of scientific qualifications amongst boy O leavers they were quite likely to continue their education and particularly their scientific education. Half of all boy O leavers enrolled for a further education course (whether full- or part-time), and 42 per cent of boy O leavers enrolled for science-based FE courses.

However, only a quarter of girl O leavers enrolled for an FE course, and only 3 per cent of girl O leavers enrolled for science-based FE courses. These girls tended to concentrate in paramedical courses, leaving pure science, engineering and technology, and agriculture as almost exclusive male preserves at this level. This imbalance in the number of girls and boys taking science-based FE courses cannot be attributed solely to girls' poor qualifications in science, since nearly half the boys entering such courses had attempted no science O grades. The boys on science-based FE courses were slightly better qualified than the girls on these courses in terms of the mean number of O grade sciences (i.e. physics, chemistry or biology) they had attempted, but, on average, neither girls nor boys were at all well prepared.

The situation was similar in many respects among H leavers. Even at this level there was a substantial proportion of pupils (38 per cent of girls and 19 per cent of boys) who had attempted no science O grades. This cannot be attributed to lack of ability in this group, but must indicate that the pupils, their parents or their teachers did not consider science to be an esential part of education, especially for girls. Even higher proportions (63 per cent of girls and 34 per cent of boys) attempted no science H grades in the fifth year. Approximately equal proportions of girl and boy H leavers enrolled for FE courses when they left school (roughly one-third) but the boys were nearly three times as likely as the girls to enrol for science-based courses (17 and 6 per cent of H leavers respectively). Among H leavers, however, unlike the O leavers, the girls who started science-based FE courses were (on average) as well qualified in science as the boys. The same was true for science-based degree courses: many more boys than girls enrolled for these courses (23 per cent of boy H leavers as against 9 per cent of girl H leavers), but the girls who did enrol were as well qualified as the boys.

Table 9.3 shows the way decisions about examination subjects are related to subsequent tertiary courses. It is clear that, for H leavers at least, the decisions about O grade subjects are extremely important. Table 9.3(a) is concerned with leavers who possess a minumum general university entrance qualification.[5] Among these leavers only 2 per cent of the girls and 7 per cent of the boys who had attempted less than two science subjects at O grade eventually read for science-based degrees. Even if they eventually left school with high general qualifications it was, therefore, highly improbable that pupils who had attempted no science O grades, or only one, would study science at university. In most cases these pupils would have decided on their O grade subjects at the end of the second year, that is, at fourteen years of age. At this early age they had effectively cut themselves off from science. This might not be serious if it affected only a small proportion of pupils. But such was far from being the case. Two-thirds of these extremely able girls had attempted only one science O grade or none at all. The same was true for nearly half the boys. The progression from H grade to degree subject was almost as predictable, with only 3 per cent of the girls and 9 per cent of the boys with minimum university entrance qualifications who took less than two science subjects at H grade in the fifth year going on to science degrees (mainly in mathematics). Pupils who had attempted two or three O or H grades in science were far more likely to study science at university, but even within this group fewer girls than boys continued their scientific education.

Table 9.3(b) shows the relationships between SCE attempts and tertiary courses for pupils entering FE colleges. I have chosen a lower 'qualification' level for these leavers with a division between 'attempting one or more science O or H grades' and 'attempting no science subjects'. For the O leavers, presumably enrolling for lower-level FE courses, sex was a far more important determinant of type of course than was qualification. Over 80 per cent of the boys and less than 20 per cent of the girls enrolled for science-based courses, irrespective of the number of science O grades they had attempted. Highers (H) leavers entering further education exhibited a pattern midway between that of H leavers starting degree courses and O leavers entering further education. In this group, pupils who had taken at least one science subject for O or H grade were considerably more likely than those who had taken no science subjects to begin a science-based course. But, at every level of qualification, boys were more likely than girls to enrol for science-based courses. Moreover, among H leavers, poorly qualified boys (those who had attempted no science subjects) were as likely as better-qualified girls (those who had attempted at least one science subject) to study science at college.

In summary, then, this analysis of science enrolments in Scotland has shown that more boys than girls study science at all levels of the educational system. In itself this is not an unexpected result, but some of the more detailed figures are both unexpected and worrying. Nearly half of all the girls who left school in 1975–76 without attempting Highers had dropped science completely by the end of S2. Over three-quarters of the girls who left school with some O grades but no H grades (O leavers) had not attempted any science O

Table 9.3 The relationship between attempts in science O and H grades and later specialisation

(a) *Leavers with minimum university entrance qualifications*[a]	*% of those with MUEQ*[a] *who had attempted given number of science subjects*			*% of those with MUEQ*[a] *and given number of science attempts who read for degree in science-based subjects*[b]		
	Girls		*Boys*	*Girls*		*Boys*
H grade attempts						
Two or three sciences in S5	19	**	43	44	**	54
None or one science in S5	81	**	57	3	**	9
O grade attempts						
Two or three sciences	33	**	57	28	**	44
None or one science	67	**	43	2	**	7
N	3,912		3,393			

(b) *Leavers entering FE*	*% of those entering FE who had attempted given number of science subjects*			*% of those entering FE with given number of science attempts who studied science-based subjects*[b]		
	Girls		*Boys*	*Girls*		*Boys*
H leavers						
H grade attempts						
At least one science in S5	33	**	60	32	**	64
No science in S5	67	**	40	11	**	45
O grade attempts						
At least one science	64	**	88	24	**	60
No science	36	**	12	8	**	30
N	1,610		1,283			
O leavers						
O grade attempts						
At least one science	26	**	53	19	**	85
No science	74	**	47	12	**	83
N	1,153		2,196			

Notes

a MUEQ is a minimum university entrance qualification, arbitrarily defined as at least four H grades, including at least one grade A or two grade Bs.

b Science-based subjects are defined as pure science, engineering and technology, medicine and agriculture.

** Sex difference significant beyond the 1% level.

grades. Two-thirds of the girls who eventually obtained a minimum university entrance qualification had attempted only one if any science O grades.

Educational decisions are not irrevocable, but these figures have also shown that a large amount of 'channelling' takes place early on in school. By this I mean that, once pupils have taken the first decision to drop science, they are set, whether they know it or not, upon a non-scientific course. This applies particularly to the more able pupils. Under present school and university arrangements, pupils who had attempted less than two science O grades would find it difficult to switch to degree specialisation in science if they wished to do so. This early channelling affects a large proportion of the more highly qualified pupils, and especially girls. But early decisions are not wholly responsible for girls' under representation in science. At every qualification level and every transition point, boys are more likely than equally qualified girls to continue with science.

Comparisons with England

Unfortunately detailed information on the uptake of science subjects at varying ability levels and the degree to which later educational decisions are structured by early choices is not available for England. However the figures presented in Chapter 1 (Table 1.1, column (c)) show that, although exact comparison is not possible, the pattern of boys' and girls' uptake of science subjects is similar in England. Many more boys than girls study physics, somewhat more boys than girls study chemistry, and more girls than boys study biology in both educational systems (although the proportions of school leavers studying these subjects differ slightly). Comparisons of the way early choices structure later specialisation among able pupils in England and Scotland also show great similarities in the extent and severity of channelling (Kelly, 1976). Pupils studying less than two science subjects at O level in the 1960s were most unlikely to take science subjects at A level or university; and over 60 per cent of the able girls who eventually obtained two A levels had passed less than two science O levels (McCreath, 1969; Phillips, 1970). Although Scottish pupils study more subjects, on average, than English pupils in the last years of secondary school, this does not mean that the vital choices are delayed. Similar proportions of school leavers with minimum university entrance qualifications in England and Scotland have studied sufficient science to continue it at university (Kelly, 1976). There are no data on less able pupils, or

(a) Pupils who continued at least one science subject into S3	Percentage agreeing with statement			
	H leavers		O leavers	
	Girls	Boys	Girls	Boys
Why did you do this subject (or these subjects) in your third year?				
I had to take the subject(s)	15 **	23	28	27
I liked the subject(s)	67	61	46	40
I was good at the subject(s)	36 **	48	12	15
The subject(s) mixed well with the other subjects I chose	33	39	17 **	29
I wanted to specialise in this area	21 *	27	14	8
Teacher(s) advised me to do the subject(s)	32	37	16 *	25
Could be useful for job or further education	55	60	32 *	41
The subject(s) were well taught	17	16	18	13
Do you consider the subject(s) to have been ... (% responding 'yes')				
Worth your while to study?	79	83	43	47
Helpful when you applied for jobs?[a]	56	63	15 **	36
Helpful when you applied for further education[a]	73	77	23 **	39
N	337	456	196	343

(b) Pupils who dropped all science subjects before S3	Percentage agreeing with statement			
	H leavers		O leavers	
	Girls	Boys	Girls	Boys
Why did you not do any of these subjects in your third year?				
Subject(s) not offered to me by my school	1	2	11	14
Not possible to take subject(s) with another subject I wanted to take	56	48	42 **	27
I didn't like the subject(s)	50	61	50	53
I found them difficult	29	34	26	30
They didn't mix well with other subjects I chose	45	37	25 *	17
Not useful for job or further education	26	22	31 **	19
My school didn't think them important	0	0	1	3
They were badly taught	4	5	3	5
Have you ever regretted not doing any of these subjects? (% responding 'yes')	30	30	19	24
Do you think any of these subjects might have helped you (% responding 'yes')				
When you applied for the jobs?[a]	30	38	16 **	36
When you applied for further education?[a]	27	33	26	28
N	205	64	373	182

Notes

a percentage of pupils who actually applied.
* Sex difference significant beyond the 5% level.
** Sex difference significant beyond the 1% level.

on the qualifications of school leavers entering science-based courses in further education in England, but where comparisons are possible the similarities between the two countries are more striking than the difference.

Pupils' rationale and opinions

The questionnaires sent to the Scottish school leavers suggested a series of possible reasons for dropping or continuing science in the third year. Respondents were asked to indicate those with which they agreed. These reasons, and the pupils' opinions about science, may help to explain the patterns of science enrolments. The results are shown in Table 9.4.

The most striking feature of this Table is that girls and boys gave remarkably similar reasons for their subject choices. For both sexes and both groups of leavers, one of the most common reasons for continuing science was simply liking the subject, while one of the most common reasons for dropping science was not liking the subject. This moves the problem one stage further back: why do so many more boys than girls like science? Another common reason for continuing science was its perceived usefulness for jobs and further education, and this was mentioned more often by boys than by girls, especially among O leavers. This accords with the previous results showing that boys are more likely than girls to enrol for science-based courses when they leave school. A substantial proportion of pupils dropped science because it was not possible to take it with another subject which they wanted to take, and this was more common for girls than for boys. So it may be that the way the option systems in schools are structured discourages girls from science. More girls than boys said they dropped science because it didn't mix well with their other subjects, whereas more boys than girls said they took science because it did mix well with their other subjects. In addition, more boys than girls among the H leavers were not given the opportunity to drop science. These results suggest that timetabling is important, and that science is more often timetabled against subjects traditionally taken by girls than against subjects traditionally taken by boys.[6] Boys appeared to be more influenced than girls by their success in science—they were more likely to say they continued it because they were good at it, and more likely to say they dropped it because it was difficult. Whether or not the pupils thought the subject was well taught did not seem to be important in decisions to drop or continue science.

Pupils were also asked to agree or disagree with some general statements about science. Table 9.5 shows the sex differences in the responses to these items. The most striking difference concerns job prospects. Many more boys than girls agreed that science is useful for getting a job, and this was true in all ability groups and for pupils who dropped science as well as those who continued it. Again this may accord with reality in that science is more useful in getting the sort of jobs that boys get than the sort of jobs that girls get, and pupils' responses are conditioned by their stereotyped expectations. Except among the best qualified pupils, boys were also more likely than girls to agree that science is necessary for a balanced education. However, by and large, girls' and boys' responses were again remarkably similar. They were also extremely positive. Around 80 per cent of all pupils agreed that science is interesting, and that practical work in the laboratory is enjoyable. It is hard to reconcile this with the large proportion of girls in Table 9.4 who dropped science because they didn't like it! Perhaps it was some other aspect of science which they disliked. This is discussed below.

The pupils were generally agreed on the difficulty of science. Almost half the best qualified pupils and over 70 per cent of the least qualified agreed that it was difficult.[7] When examined in detail it appears that the scientific concepts were the source of the difficulty for well qualified pupils, but that for less successful pupils the vocabulary and the use of maths were as troublesome as the ideas. These results should certainly give teachers and curriculum developers cause for thought. There was a consistent trend for girls to admit to greater difficulty than boys with science, but this trend was only significant in a few cases.

Perhaps the most interesting point to emerge from this analysis of pupils' rationale for their science choices concerns the perceived relevance of science to pupils' future lives. Girls were far less likely than boys to agree that science is useful for getting a job. This is realistic in terms of the traditional job market for girls and boys. But it suggests that girls are not thinking beyond this traditional job market and are limiting their career aspirations to stereotypical women's jobs which do not require scientific training.

Some further evidence for the differing connotations of science for girls' and boys' careers comes from a small survey carried out by Judy Samuel (private communication). She asked pupils taking chemistry in the fourth and fifth year of an English comprehensive school what was their main reason for studying chemistry, which other sciences they were taking and what if any career plans they

Table 9.5 Percentage of girls and boys agreeing with various statements about science by presentation level and when they last studied science

	Presentation level								H leavers				O leavers			
	Attempted 5+ H grades in S5		Attempted 1-4 H grades in S5		Attempted 6+ O grades but no H grades at school		Attempted 1-5 O grades but no H grades at school		Continued some science into S3		Dropped all science before S3		Continued some science into S3		Dropped all science before S3	
	Girls	Boys	Girls	Boys	Girls	Boys	Girls	Boys	Girls	Boys	Girls	Boys	Girls	Boys	Girls	Boys
'Here are some things people think about science (physics, chemistry and biology). For each statement, tick one box to show whether you agree or disagree with it.' (% agreeing)																
'Science is necessary for a balanced education'	76	78	52 **	65	36 **	63	31 **	53	70	73	35	39	39 **	60	25 **	43
'Science is interesting'	86	91	79 **	90	76 *	86	80	82	89	92	64	67	84	87	72	73
'Practical work in the laboratory is enjoyable'	81	85	85 *	91	88	92	86	87	87	90	75	78	91	92	83	80
'Science is useful for getting a job'	60 **	80	50 **	69	33 **	63	31 **	48	58 **	75	42	56	33 **	53	31 **	52
'Science is difficult'	48	44	69 **	57	73	65	74	70	60 *	52	72	54	71	69	76 *	64
'Science involves too much maths'	18	19	49 *	38	49	48	55	59	34 *	28	53	54	47	54	60	60
'In science lessons you can find things out for yourself'	n.a.		n.a.		79	81	82	87	n.a.		n.a.		83	85	79	84
'Science helps you to understand things in everyday life'	n.a.		n.a.		74	78	77	74	n.a.		n.a.		77	75	74	76
'When you last did science, were any of the following things problems for you?' (% responding 'yes')																
'The words that were used'	9	9	17	18	40	31	51 *	44	14	14	20	18	48 **	38	48	44
'The maths that was involved'	22	22	48	42	59	50	67	66	40 *	32	45	53	62	60	66	61
'The ideas which had to be understood'	45 *	31	52	46	56	50	56	54	46	41	62 **	32	56	52	56	54
N	149	189	376	303	196	230	406	335	369	450	162	46	329	454	288	136

Notes

* Sex difference significant beyond the 5% level.

** Sex difference significant beyond the 1% level.

n.a. Not applicable; H. leavers were not asked to agree or disagree with these statements.

had. Several of the boys said they were studying chemistry because they would need it in their careers, when in fact the careers they had in mind did not require qualifications in science (e.g. accountant, banking, law, town and country planning) or required some science but not chemistry (e.g. electronics, computer programming, mechanical engineering). Conversely several of the girls were considering careers, such as nursing and catering, for which chemistry would be a useful qualification but did not give their career plans as a reason for choosing chemistry. Although this study confirmed that boys were more likely than girls to be considering science-based careers, it also suggested that boys were *over*estimating the usefulness of science for their careers, whereas girls were *under*estimating its usefulness. Further analysis of the data confirmed this impression. When the pupils were divided into three groups, according to their career plans (scientific, non-scientific or non-existent), it emerged that approximately 60 per cent of the boys taking chemistry were also taking physics and biology, irrespective of which group they were in. Boys planning non-scientific careers or with no career plans were frequently keeping their options open by studying all three sciences. But less than 20 per cent of the girls in these two groups who were taking chemistry were also taking physics and biology. Although the girls who were planning scientific careers were as likely as the boys to be studying three sciences, girls who were undecided or who currently intended non-scientific careers had frequently dropped physics and so effectively eliminated the possibility of deciding on a science based career at a later stage. In many cases this appeared to be as a result of advice from members of staff who perhaps did not consider science as important for girls as for boys.

In the Scottish study most of the reasons for choosing science and opinions about science were endorsed to a similar extent by girls and boys. The sex differences in the responses were minor compared to the sex differences in science enrolments. This suggests that these questions were not tapping the major factors which distinguish between girls' and boys' attitudes to science. At present we can only make an informed guess at the nature of these major factors, but an informed guess does provide a starting point for further investigations.

I argued in Chapter 5 that the masculine image of science may be the main factor which deters girls from enrolling in science courses. This explanation can account for the responses to the questions on reasons and opinions discussed above. Perhaps girls say they dislike science, despite finding it interesting and enjoyable, because they dislike being involved in a boys' subject. When science is competing on the timetable with other subjects which have a feminine image, girls may prefer these other subjects. Girls more often than boys mentioned the impossibility of taking science with another subject they wanted to take, and this clash suggests that the school authorities too may see science as a boys' subject and place it in opposition to girls' subjects on the timetable.

The result is that pupils are channelled towards or away from science-based courses and careers on the basis of their sex. For H leavers this channelling takes place subtly, through the medium of subject choice at fourteen. Most girls take little or no science at this stage, and so lack the qualifications to proceed to more advanced study in science. Later selection is less obviously sex-dependent, since qualified girls continue in science to almost the same extent as qualified boys. But the major sorting by sex has already been made before the third year. Among O leavers boys also take far more science than girls, but the channelling is more crudely by sex. So pupils are slotted into their position in the labour market, scientific or non-scientific, higher or lower, according to their sex and academic qualifications.

I am not implying by this any conscious attempt by teachers to direct girls away from science (although there is evidence of hostility towards girls in science from some teachers: see Chapter 17). Teachers operate the system, and the system operates against girls in science. Nor am I implying that girls feel constrained by their narrowing options. Quite the reverse: girls often abandon science willingly and eagerly. But they do so against a background of sex stereotyped assumptions and expectations which renders the notion of 'free choice' largely meaningless. The channelling is apparently voluntary but that does not make it any the less deplorable.[8]

Notes

1 This paper first appeared in a slightly different form in the *Collaborative Research Newsletter* of the Centre for Educational Sociology, University of Edinburgh, Nos. 3 and 4.

2 This unique facility consists of a series of sample surveys of Scottish school leavers from 1962 to 1977. The data are stored on a computer in a readily accessible form, and are made available to all interested parties, be they teachers, administrators, parents or educational researchers. Introductory courses on the data and the computing system are held from time to time, and the Collaborative Research team will help in

preparing and running computer programs. My own programs were submitted by post from Manchester, and I am extremely grateful to Andrew McPherson and John Gray for the time and effort they put into running them. Further details of the archive and its use are available from the Centre for Educational Sociology, University of Edinburgh.

3 The Scottish education system is distinct from the English system, and has its own examinations. Although the normal age for starting school (five) and the minimum school-leaving age (sixteen) are the same as in England, pupils transfer to secondary school a year later at the age of twelve. O grades are usually attempted at the end of the fourth year (S4) at the age of sixteen. There is no equivalent of the CSE examination, but the Scottish O grade is at a slightly lower level than the English O level, and at least one subject is attempted by over 70 per cent of the age group (*cf.* over 50 per cent of the age group attempt at least one O level in England). H grades are usually attempted for the first time in fifth year (S5) and frequently repeated, in the same or different subjects, in sixth year (S6). H grades are generally agreed to be at a standard between that of the English O and A levels. They constitute the entrance qualification for higher education, and pupils aiming for university usually take about six or seven subjects. For further details of the Scottish education system see Osborne (1966, 1968).

4 Because of the rotating design of this part of the questionnaire, the number of pupils considered in this chapter varies from table to table. The rotating design also means that the reader would be unwise to 'rework' the raw numbers in any table to reach conclusions other than those that the table explicitly intends. This applies especially to comparisons between the H, O and non-certificate levels. Percentages, however, can be compared; indeed, that is the whole purpose of the tables.

5 Although Scottish universities will in theory accept students with three H grades, entrance requirements are more normally set at four Highers with good grades in at least two of them. The honours degree course in Scottish universities is usually four years long but applicants with good A levels can be exempted from the first year.

6 *Cf.* Ormerod's suggestion (Chapter 7) that the option system accords better with boys' subject preferences than with girls'.

7 It is possible that they wold have said the same of any other subject had they been asked—which they were not.

8 I would like to thank Judy Samuel for reading and commenting on an earlier draft of this chapter.

References

Kelly, A. (1976), 'Swings and roundabouts: trends in SCE science presentations since 1962', *Scottish Educational Studies*, Vol. 8, 4.

McCreath, M. D. (1969), 'Curriculum patterns in the O level base for sixth form work', Royal Statistical Society.

Osborne, G. S. (1966), *Scottish and English Schools*, Longman.

—(1968), *Change in Scottish Education*, Longman.

Phillips, C. M. (1970), 'Some changes in the factors affecting university entry', *Research in Education*, Vol. 4, 81.

10

Who says girls can't be engineers?

PEGGY NEWTON

'There is something not quite natural about a girl wanting to be an engineer,' remarked a supervisor in a large engineering company in the West Midlands. His comment, echoed by many of his colleagues, reflects the strong masculine image of engineering in Britain today, where only 2 per cent of jobs at technician level or above are held by women (EITB Annual Report, 1977–78). Although the Sex Discrimination Act became law in 1975, ideas of 'careers for boys' and 'jobs for girls' remain deeply entrenched in many schools and Careers Services, as well as in industry. Girls wanting to take technical drawing or metalwork at school—if not met by outright opposition from the head teacher—often receive the message from teachers and classmates that these are not suitable subjects for girls.

A recent attempt to increase the number of women at technician level in engineering highlights some of the problems facing girls who decide to choose engineering as a career. In the autumn of 1976 the Engineering Industry Training Board (EITB) began an experimental programme to recruit and sponsor approximately fifty girls each year in their first two years of training as technicians. The information reported in this chapter is taken from structured individual interviews with girls on the EITB programme who began their training in 1977 and 1978.[1] The girls' comments indicate that many of them had not been aware, while they were at school, that engineering was a possible career choice. Those who were seriously interested in engineering had frequently had difficulty in gaining admission to secondary school subjects that are usually considered necessary for technical careers. Their comments also draw attention to biases and shortcomings in many schools careers programmes and in the Careers Service.

Although there are wide variations within the industry, a

technician is expected to understand basic engineering skills and processes and to be able to translate instructions given by professional engineers to craftsmen or craftswomen on the shop floor. Technicians are employed in a variety of areas, including inspection and testing, research and development, and work study and maintenance. Entrants to technician training are usually sixteen -year-old school leavers who have at least three or four examination passes (either O level or CSE) in subjects related to engineering.

However, in some companies technicians have originally been trained at a craft level and are subsequently 'promoted' to technician status. Most technician trainees undergo four years of study and training and work for a certificate awarded by the Technician Education Council. These qualifications are replacing the City and Guilds and Ordinary and Higher National Certificates and Diplomas in various fields of engineering.

The girls who were recruited to the EITB scheme came from two regions of the country—the West Midlands and Surrey.[2] During their first year the girls were trained in all female groups, ranging in size from seven to fourteen. Several different patterns of training were used, with girls being based either in a company, an EITB regional training centre or a college of further education. Girls who were placed in a company or a training centre also attended college on a day release or block release system. In all the training sites girls were segregated from boys, although in many cases the training was similar to or identical with that received by the boys. In their second year of training girls in the West Midlands were placed in groups of three or four in companies sponsored by the EITB and returned to college for additional study. In Surrey the girls also combined industrial experience with college work but girls were placed in companies singly or in pairs. Although the girls were recruited from the same area, they had attended a wide variety of secondary schools, ranging from small independent boarding schools to large inner city comprehensives. Very few of the girls in the first year's intake had been to the same school, but it was noticeable that in subsequent years girls were more likely to be recruited from schools which had a former pupil already in the scheme.

Research strategy

To provide comparisons with the groups of girls in engineering, several control groups were selected. The first group, which will be designated '1977 boys in engineering', consisted of boys who were being trained as technicians on the same sites as the girls. The second group, which will be referred to as '1977 girls holding or preparing for traditional jobs' comprised three sub-groups, which have been combined for the present discussion. Two sub-groups represent girls chosen because they were doing two-year college courses which prepared them for traditionally feminine careers—nursery nursing and secretarial work with a business studies qualification. The final sub-group comprised girls who were friends and secondary school classmates of the girl engineers and who were either working in or preparing for traditionally feminine jobs.[3]

Subject choice

Since most employers and colleges offering courses in engineering strongly recommend that applicants have examination passes in physics and mathematics and some experience of crafts or technical subjects, it is important to look at the availability of these subjects to girls. Although almost all the girls in engineering had taken mathematics for either an O level or a CSE exam, physics was a more unusual choice and several girls reported having been the only girl in the class at school. Two schools did not allow girls to take physics, and some girls described having to 'put up a fight' to be allowed to do the subject.

Girls were frequently unable to take technical or craft subjects, such as woodwork, metalwork or technical drawing. Of the ninety-one girls in engineering (1977 and 1978 groups combined) who answered questions about subject choice, forty-eight said they had wanted to take technical subjects at school but were not able to do so or were not allowed to take the subjects on the same basis as the boys. As can be seen in Table 10.1, the most frequent reason for not being able to take technical subjects was that there were no facilities at the school. In nineteen out of the twenty-one cases, facilities were not provided because the girls attended a single-sex school. However, two of these girls noted that arrangements had been made for boys to take cookery at their schools, but they were not allowed to take technical or craft subjects at the boys' school. In over a quarter of the cases, girls attended schools in which girls were not allowed to take technical subjects or schools which operated a policy of giving boys the first places in technical subjects. The 'boys first' policy is usually justified by the assumptions that boys need these subjects for apprenticeships but that the girls only do the subjects for amusement or to attract the attention of the boys in the class. It operates effectively to bar girls from the subjects or to see that only one or two girls are allowed in any class, thereby preserving the areas as 'boys' subjects'.

Table 10.1 Reasons given by girls who eventually went into engineering for not taking technical subjects at school

Reason	*N*	%
No facilities	21	44
Girls not allowed or 'boys get first places'	13	27
Girls officially allowed but strongly discouraged by teachers and/or not permitted to take exams	5	10
Timetable conflicts	5	10
Not socially accepted for girls to do technical subjects	3	6
Parent opposed	1	2

Five girls described experiences of unequal treatment. Two were strongly discouraged from taking technical subjects by the subject teacher and three were allowed to take the subjects but were not permitted to sit the examinations at the end of the course.

Another reason for not doing technical subjects was social pressure. Although many of the girls who had done technical subjects at school proudly described struggles to be allowed a place, others were unwilling to 'create a fuss' and were uncertain about being the first girl in the school to do a 'boys' subject'.

Although the Sex Discrimination Act was intended to eliminate differences in educational opportunity, many schools appeared reluctant to change traditional policies. From the girls' spontaneous comments about subject choice it seemed that very few schools automatically allowed girls to do technical subjects. In most cases girls and their parents had to appeal to the head teacher to change school policy. Two schools attended by the girl trainees have been threatened by parents of younger girls with court action over subject choice. At one school girls are now allowed a free choice of subjects, but court action is still pending in the second case. Since single-sex schools are not subject to the portions of the Sex Discrimination Act regarding educational opportunities, girls attending these schools have no way of compelling their schools to offer technical subjects.

Several girls who had not been able to do technical subjects reported that their raising of the issue had made it possible for girls in lower years to do the subject. Four girls took courses at night school or at a college of further education when they were not able to do the subjects they wanted at school.

If the technical subjects taken by girls and boys entering engineering in 1977 are compared, an interesting pattern emerges. Not surprisingly, more boys (77 per cent) than girls (30 per cent) had taken one or more technical subjects. However, of the thirty-seven boys who had taken technical subjects, twenty-nine had taken technical drawing or design, twenty had taken woodwork or metalwork and ten had taken courses in some type of engineering. Of the fourteen girls who had taken technical subjects, all had done technical drawing and four had also done metalwork. No girls had taken woodwork or engineering science, although several girls had wanted to take these options. These results suggest that girls are most likely to be allowed to take technical drawing in schools but that other craft subjects remain a male preserve. The trend fits with the sentiment often expressed by men in engineering that 'if women are going to be engineers, they are best suited for jobs in the drawing office'.

When girls and boys were asked if there were '. . . any options that you wanted that you weren't able to get at school', boys in engineering were far more likely to be pleased with their selections than girls in engineering or girls in traditional fields. Boys and girls who were dissatisfied with their subject choices were asked to name subjects they would have preferred. If these subjects are classified as 'career-related' or 'other', a majority of respondents in all groups

Table 10.2 Comparison of groups' satisfaction with subjects taken for examination

Group	*No.*	*% satisfied with subject choices ***	*% of total desiring more career-related subjects*
1978 girls doing engineering	44	36	61
1977 girls doing engineering	47	30	45
1977 boys doing engineering	48	71	21
1977 girls holding or preparing for traditional jobs	64	41	45

Note

** Difference between the groups significant beyond the 1% level (χ^2 test). (See the Statistical Appendix for a discussion of significance testing and the χ^2 test.)

wished they had taken more career-related subjects. If these numbers are expressed as percentages of the total numbers in each group, differences among the groups are readily apparent. Only 21 per cent of the boys in engineering wanted more career-related options, but 61 per cent of the 1978 girls in engineering, 45 per cent of the 1977 girls in engineering and 45 per cent of the girls in traditional fields wanted additional career-related subjects (Table 10.2).

Entry into engineering

Given the difficulties girls face in doing technical subjects in school and the strongly masculine image of engineering, it is useful to consider how girls heard about the EITB course. Girls were most likely to find out about engineering from school or from the Careers Service (see Table 10.3). Over half the sample utilised these sources of information. In contrast, only 31 per cent of boys in engineering and 33 per cent of girls holding or preparing for traditional jobs made similar use of the school or the Careers Service. The

Table 10.3 Sources of information about jobs and courses: comparisons of boys in engineering, girls in engineering and girls holding or training for traditionally feminine jobs (%)

Source of information	*Boys in engineering*	*Girls in engineering*	*Girls holding or training for traditional jobs*
Newspaper or radio***	31	34	4
Careers teachers from school or the Careers Service*	31	51	33
Friends, parents, 'contacts', etc.***	27	10	39
Self, i.e. wrote or phoned for information	4	0	16
A college of further education	4	2	3
Other	2	2	4
N	45	90	67

Note
* Differences between the groups significant beyond the 5% level (χ^2 test).
*** Differences between the groups significant beyond the 0·1% level (χ^2 test).

differences between the groups in the use they made of the school and Careers Service are statistically significant beyond the 2 per cent level.

When girls in the 1978 sample were asked when they first considered engineering as a career, fifteen out of the forty-four said they had not thought about engineering until they heard about the course. However, most of the girls had considered engineering during their final two years at school at a time when it was too late for them to change their choice of subjects for examination. Eight of the girls had seriously considered engineering before or during their third year at school.

Several of the girls who first considered engineering when they heard about the course said that they hadn't known that the field was open to women. One girl who decided on engineering at age eleven remembers a television programme showing a woman engineer which started her thinking about her own future. Such anecdotal evidence suggests that increased public visibility of women in engineering is required if the numbers of women in the field are to be increased.

Role of school careers and the Careers Service

School careers programmes and the Careers Service played a major role in over half the girls' decisions to enter engineering, but girls' opinions of the advice provided varied widely. When asked to rate careers advice on a five-point scale, girls choosing engineering and girls selecting traditional careers were more likely than boys in engineering to use neutral or negative terms to describe the careers advice they had received (see Table 10.4). Girls in engineering were especially prone to feel strongly negative, with 33 per cent of the sample labelling school careers advice as 'a complete waste of time'. If the mean ratings of the school careers service made by girls in engineering and boys in engineering are compared, the difference between the two groups is significant beyond the 5 per cent level.

Detailed and systematic information about the provision of careers lessons at school is available only for the 1978 sample of girl engineers. It suggests a wide variety in the type of careers advice offered by the schools the girls attended. Out of the forty-four girls interviewed, twelve reported having had no careers lessons or advice at school. In two instances, careers advice was available but the girls chose not to make use of it. Among the girls who had careers lessons at school there was no relationship between their ratings of school careers advice and the quantity and variety of careers information provided.

Looking at the comments of the nine girls interviewed who rated their experience of careers advice at school as 'very useful' or 'quite useful', almost all described careers programmes which covered a wide range of jobs and teachers who appeared to leave the choice and initiative clearly with the student. Many said that the careers teacher had helped them to find a general area and told them about the sorts of jobs and opportunities available. Five of the nine girls rating school careers advice positively heard about the EITB training programme from a careers teacher or subject teacher at school. Although hearing about engineering from the school appeared to contribute to a positive rating of careers advice, it was not a sufficient condition. In the group of girls who rated careers advice at school as 'not at all useful' or 'a complete waste of time', four had learned about the EITB scheme at school.

Table 10.4 Comparison of groups' ratings of careers advice at school (%)

Rating	*Boys in engineering*	*Girls in engineering*	*Girls holding or training for traditional jobs*
Very useful	16	13	6
Quite useful	36	20	24
Somewhat useful	13	24	31
Not at all useful	20	10	22
A complete waste of time	16	33	16
N	45	76	67

Note
The differences between girls and boys in engineering were significant beyond the 5% level (two-tailed Mann–Whitney U test).

Complaints about careers teaching at school were surprisingly vague. A few girls were critical of teachers or schools who had only one or two lines of advice. They described schools in which most pupils were expected to go to university or at least to A levels and the possibility of leaving school at sixteen was rarely considered. They also mentioned teachers who had conventional stereotypes of women's work roles, advising all girls to be nurses or secretaries.

Of the twenty-six girls attending mixed schools and having

careers lessons, eleven reported that boys and girls were always together for careers lessons. However, in eight schools boys and girls were almost always or always separate and frequently received different information about career opportunities. In seven schools boys and girls were usually together for careers teaching. Not surprisingly, careers lessons were most frequently separate in these schools when a career topic was assumed to be of interest only to one sex. For example, in one school only girls were included in careers talks on shop work and hairdressing and only boys attended talks on engineering. In other schools all careers talks were theoretically open to both sexes, but it was expected that only one sex would attend, so that boys going to talks on nursing or girls going to discussions of engineering were likely to feel strange and out of place.

Girls attending single-sex schools were somewhat less likely than girls at mixed schools to have had careers lessons; 87 per cent of the girls attending mixed schools had had careers lessons and 57 per cent of the girls from single-sex schools had received careers lessons. However, the difference between the two goups is not statistically significant for a sample of this size. If the girls' ratings of the careers advice are compared, there is no difference between the two types of school, nor is there a difference in the proportion of girls encouraged in engineering by their schools.

There were no significant differences among the groups (girl engineers, boy engineers and traditional girls) in their ratings of the Careers Service (see Table 10.5). Boys' ratings of the Careers Service were positively correlated with their ratings of school careers advice ($r = 0{\cdot}40$), as were the ratings of the girls in traditional fields ($r = 0{\cdot}44$); however, there was no relation between the two sets of

Table 10.5 Comparisons of groups' ratings of careers service (%)

Rating	*Boys in engineering*	*Girls in engineering*	*Girls holding or training for traditional jobs*
Very useful	24	24	10
Quite useful	34	30	27
Somewhat useful	17	21	37
Not at all useful	24	25	27
N	29	76	41

ratings for the girls in engineering.

If the comments of the 1978 girl engineers on the Careers Service are examined, several themes emerge. Seven of the nine girls interviewed who rated the service as 'very useful' justified their ratings with a statement about the careers adviser having told them about the EITB training programme. This reason was also given by four of the thirteen girls who described the service as 'quite useful'. Six of the girls who described the Careers Service as 'very useful' or 'quite useful' mentioned the sympathetic attitude of the careers advisers they met. 'They were very helpful and seemed to understand how we felt', said one girl. Another praised her careers adviser as being 'someone to talk to and discuss thing with. You could go in with an idea and get their opinion on it'.

The girls who felt that the Careers Service was 'not at all useful' most often remarked on advisers trying to push them into jobs they didn't want to do. One girl complained, 'We knew what we wanted to do and they had a fixed job for us to go to. They tried to push us . . .' The positive side of this sort of comment can be seen in the statement of a girl who rated the service as 'quite useful'. 'She took more interest and didn't force me and she listened to what I wanted to do—not what she wanted to do.'

Another common reason for rating the Careers Service as 'not at all useful' was that a girl didn't know what she wanted to do. This is obviously a difficult situation for both the girl and the adviser, and unless one of them produces an idea which both find acceptable it is likely to end in an unsatisfactory consultation.

When the girls' comments on school careers lessons are compared with their comments on the Careers Service it appears that the girls have different expectations about the type of advice that is appropriate. At school they seem to expect and favour a fairly broad general approach which gives them ideas about a variety of jobs. When they see the careers adviser (frequently after they have left school without a job), they tend to want much more specific information about available jobs and training schemes. In many situations the school careers programme and Careers Service perform complementary functions. However, over a quarter of the 1978 sample of girls in engineering had only one source of careers information, with twelve not having had careers advice at school and two not having reported any contact with the Careers Service.

Overview

Although more research is needed, the preliminary evidence reported here suggests that efforts at changing attitudes in schools, the Careers Service and industry are needed if engineering is to become more open to women. Traditional sex-role boundaries must be relaxed so that girls wanting to do metalwork are not regarded as freaks and women who choose careers in engineering are seen as sensible and competent in their work. Many of the girls in the EITB programme regard themselves as pioneers—as indeed they are—but hopefully the time is approaching when more school girls will think about engineering as a matter of course and will not create an uproar at school, as one of the EITB trainees did, when announcing, 'I am going to be an engineer'.

Notes

1 The research is supported by a Social Science Research Council Programme Grant to Professor G. M. Stephenson and by a grant from the Engineering Industry Training Board to Professor Stephenson and myself.

2 In future years it is hoped to extend the scheme to the rest of the country. The target for 1980 is to recruit 250 girls for training at technician level. For further information about the scheme contact the Engineering Careers Information Service, 53 Clarendon Road, Watford WD1 1LA

3 Although data from similar control groups were collected in 1978, they were not yet analysed at the time of writing. A more extensive interview was used with the 1978 groups, so that more specific information on subject choice and careers advice is available for the 1978 girl engineers than for the 1977 groups.

11

Predicting specialisation in science

JUDY BRADLEY

It has been traditional in our society to regard male and female roles as distinctly polarised spheres of activity. Though girls have increasingly acquired the kinds of education and training which lead to the expectation of a career, there are still relatively few of Alice Rossi's 'pioneers' who are prepared to lead the way into male-dominated occupations (Rossi, 1965). The pervasive sex stereotyping of interests and vocational preferences means that even the ablest girls are being channelled into a few socially acceptable occupations. In scientific and related professions the barriers to female recruitment have been particularly slow to fall.

Occupational choice is generally seen as a sequence of interrelated choice periods which limit or extend the range of future possibilities. One such decision-making point is in the selection of subject options in secondary school, and for girls this usually means a rejection of the physical sciences. This is not, however, true for all girls. Every year a small but significant minority choose to continue their studies in this area. By identifying the distinctive qualities of this group and locating the stage at which their subject preferences emerge we may be better able to gain an understanding of the types of intervention necessary for promoting science interest among girls on a much wider basis.

The findings reported in this chapter represent part of a longitudinal study carried out at Oxford University Department of Educational Studies. Its principal aims were to identify specific factors contributing to pupils' subject choices at each stage in their school careers and to discover how far a pattern of subject specialisation could be traced back through the early years of secondary education. In an investigation which concerns itself solely with the differences between specialist groups as they approach the end of secondary school a degree of uncertainty remains as to what extent the differences observed are a product of the various academic disciplines rather than the determinants of choice. There is evidence to suggest that existing differences are in fact accentuated by prolonged exposure to a discipline (Altmeyer, 1966; Thistlethwaite, 1973). A longitudinal study enables one to establish just how far the observed differences antedate the point at which the final decision to specialise is made. In the research to be described here data were collected at two important points prior to sixth form specialisation: in the third form when pupils are beginning to consider the possibilities open to them and in the fifth form when they are in the process of deciding on their sixth form courses.

The initial sample consisted of 1,925 boys and girls from fifteen co-educational secondary schools throughout England and Wales. They were representative of the whole ability range. During the first phase of field work these pupils were asked to indicate their favourite subject at the time. All pupils also completed an intelligence test and a personality questionnaire. Our intention was to use the scores on these tests to discover to what extent sixth form science choices could be predicted from this early information. For the second (fifth form) phase of field work there were 1,596 pupils with complete profiles remaining in the sample. They were again asked to indicate their favourite subject. 675 of these pupils stayed on to take A levels.

The results reported here concentrate specifically on a comparison of the arts and science groups of girls. For the first two stages of the research the groups were defined by pupils' choice of favourite subject. Those who chose languages or literature comprised the arts group, while those who chose physics or chemistry comprised the science group. Biology was treated separately, since this is usually regarded as more of a 'girls' subject' than either physics or chemistry and might be expected to have something of a blurring effect on the results obtained. Mathematics was also treated separately. At the sixth form stage so few girls were studying physics or chemistry without also studying biology that the definition of the science group had to be changed. All pupils who were studying any combination of physics, chemistry, biology and mathematics (but no other subject except general studies) were defined as science specialists.

A longitudinal analysis of pupils' preferences and eventual choices indicated that science specialisation is the product of an early and enduring interest in science subjects. Figure 11.1 shows how the subject preferences of the girls in our sample changed over time, whilst Figure 11.2 illustrates in more detail the early preferences of

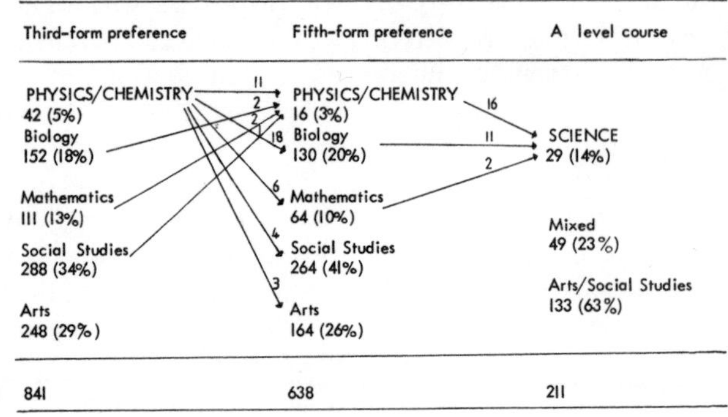

Fig. 11.1 A longitudinal analysis of the subject preferences of the girls' science groups.
Note The total number of respondents at the fifth-form stage is smaller than that at the earlier stage as, in addition to those who had left the sample schools, several girls changed their preference to non-academic subjects such as domestic science and secretarial studies. This did not, however, occur in the case of any girl expressing an early interest in physics or chemistry.

the girls who stayed on to study three science subjects in the sixth form. The pattern of subject preferences observed in the third form left no doubt that pupils were already beginning to exhibit the interests characteristically associated with their sex roles. Thirty per cent of the boys compared with only 5 per cent of the girls gave physics or chemistry as their favourite subject, while 17 per cent of the boys and 38 per cent of the girls indicated that these were the subjects they most disliked. By the time these pupils reached the fifth form physics or chemistry had remained a sustained preference for only eleven of the forty-two girls who had earlier stated one of these to be their favourite subject. A further eighteen now expressed a preference for biology, six for mathematics and the remaining seven for arts or social studies. There was only a slight shift towards physics or chemistry by girls who had originally preferred other subjects. Two came from the mathematics group, two from the biology group and one from social studies. In general, therefore, the results at both the third and fifth form stages support the suggestion that girls were using subject preferences to underline their identification with their assigned sex role. It must, however, be emphasised that this pattern was also reflected in the subject preferences expressed by the boys in the sample, among whom there was a remarkable shift in interest from arts to social studies. Almost

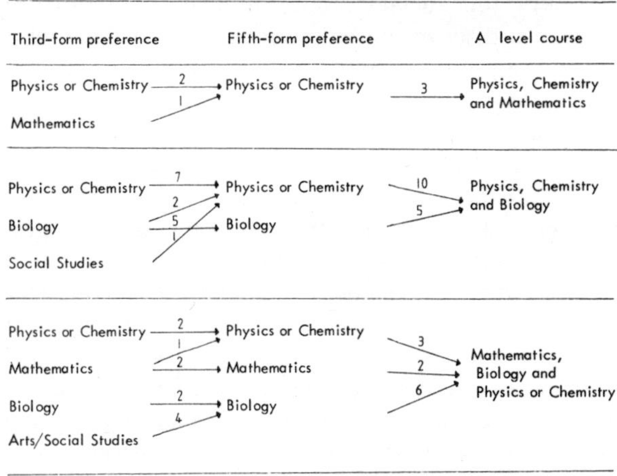

Fig. 11.2 Early preferences of the sixth-form science groups.

45 per cent of the boys who had favoured arts subjects in the third form had by the fifth form changed their allegiance to social studies, and in particular to economics. In the sciences, too, there was evidence of boys' interests shifting from biology to physics or chemistry (13 per cent).

Of the 841 girls in the initial third form sample 211 stayed on to take A levels, only seventy-eight of whom included physics, chemistry or biology among their options. Forty-four of these girls were taking only one science (usually biology) and one or two arts or social science subjects and were therefore included in the 'mixed' rather than the 'science' group. A further five were taking two science subjects, but as in each case one of these was biology these girls were also excluded from the 'science' group. The eventual sixth form sample of twenty-nine 'science girls' comprised three taking physics, chemistry and mathematics, fifteen studying physics, chemistry and biology and the remaining eleven taking mathematics and biology combined with either physics or chemistry. Thus the science specialists represented only 14 per cent of the total sample of sixth form girls. The same definition of science specialisation among the boys produced a figure of 46 per cent. These proportions provide a clear illustration of the fact that, while science is a common option for boys, girls making the same choice are taking a decisive step outside their traditional sex roles. They have embarked on a path which, in the present social climate, will lead them into areas which have acquired a distinct masculine identity (see Chapter 16). Our purpose in examining their cognitive abilities and personality characteristics was to assess how far such variables might be regarded as influential in contributing to this unusual course of action. Since the intelligence and personality profiles of females generally deviate from those associated with science specialisation, it might be expected that in both of these respects those girls who do choose to study science would be an especially interesting group.

There is mounting evidence to suggest that pupils choosing science subjects tend to attain higher scores on conventional intelligence tests than those choosing arts or social studies (Hudson, 1966; Ormerod, 1969). Science specialists, moreover, score significantly higher on items measuring spatial or diagrammatic ability, while arts specialists are more likely to be verbal in cognitive bias (Hudson, 1968; Child and Smithers, 1971). Even a cursory glance at the literature on sex differences will indicate a marked divergence in cognitive abilities between the science specialist and the female norm (Ounsted and Taylor, 1972; Maccoby and Jacklin, 1974). One of the principal tasks of this investigation was therefore seen as seeking to determine whether the evidence amassed from this earlier research—carried out predominantly with male subjects—was making an equally valid statement about academic specialisation in the female population.

The measure adopted for this aspect of the research was the AH4 Group Test of General Intelligence (Heim, 1970). This test is divided into two equal parts, performance being expressed in terms of the total test score (range 0–130). Part I consists of items with a verbal or numerical bias and Part II of items with a spatial bias.

The results presented in Table 11.1 show that the girls favouring science at each stage had averaged higher scores on the AH4 in the third form than the girls favouring arts. The mean third form scores of both arts and science students increase from stage to stage as the less able pupils leave school or change their preferences to other subjects. However, the science group's lead over the arts group is maintained into the sixth form (see also Chapter 12). The relatively greater importance of spatial ability in distinguishing between the arts and science groups is immediately apparent. Although the science girls also obtained higher scores on the verbal-numerical part of the test, this difference is significant only at the third form stage.[1] The inclusion of girls studying biology in the sixth form science group probably has no distorting effect on this analysis, since

Table 11.1 Comparison of mean third-form intelligence scores of girls favouring science and arts at successive stages

	Third form		*Fifth form*		*Sixth form*	
AH4 Test of Intelligence	*Science*	*Arts*	*Science*	*Arts*	*Science*	*Arts*
Verbal and numerical	31·84 ***	25·35	35·82	31·02	35·68	33·44
Diagrammatic	44·63 ***	35·99	46·54 *	40·77	49·18 ***	42·08
Full-scale	76·47 ***	61·34	82·36 *	71·79	84·86 ***	75·53
N	42	248	16	164	29	133

Note
*** Science–arts differences significant beyond the 0·1% level.
* Science–arts differences significant beyond the 5% level.

no significant differences were observed at the third and fifth form stages between their intelligence scores and those of the girls favouring physics or chemistry.

These findings may, then, be regarded as offering support on a wider basis for previous research concerning the role of cognitive variables in the arts–science dichotomy. The girls choosing to specialise in science did obtain higher test scores than the arts specialists and were characterised particularly by a significantly greater spatial ability. An additional analysis comparing the intelligence test scores of the girls with those of the boys in the sample found no significant differences between the test scores of boys and girls choosing science subjects at any of the three stages.

Moving on to the area of personality, we are again faced with a pattern indicating the existence of highly significant differences between scientists and the female norm. A fairly clearly defined image of the personality of scientists has been derived from a wide range of research studies (see Chapter 16). Biographical and retrospective investigations of eminent scientists have indicated confidence, ambition, single-mindedness, perseverance and an avoidance of close personal relationships and complex emotional situations as their distinguishing characteristics (Roe, 1953; Chambers, 1964; Helson, 1971). Science specialists are generally reserved, emotionally stable, conscientious, tough-minded and self-sufficient (Cattell and Drevdahl, 1955; Saville and Blinkhorn, 1976), while arts specialists are more concerned with people, more sociable and concerned with their own emotions (Smithers, 1974).

Table 11.2 Comparison of mean third-form personality scores of girls favouring science and arts at successive stages

Personality traits on HSPQ test	*Third form*			*Fifth form*			*Sixth form*		
	Science		*Arts*	*Science*		*Arts*	*Science*		*Arts*
A Sociable+ Aloof –	11·29		11·93	9·54	**	12·37	10·86		11·99
B Bright + Dull –	7·63		7·40	8·45		7·92	8·57		8·19
C Mature + Immature	8·50		7·85	9·09	**	7·76	8·18		8·08
D Excitable + Phlegmatic –	9·42		10·52	10·54		10·62	11·32		10·86
E Dominant + Submissive –	8·92	***	7·08	10·18	**	7·53	9·50	***	7·17
F Enthusiastic + Sober –	10·42		9·90	9·27		9·77	9·61		10·55
G Conscientious + Casual –	11·57		10·85	13·82	*	11·67	13·32		12·29
H Adventurous + Timid –	9·82		8·63	8·64		8·26	9·32		8·84
I Sensitive + Tough-minded –	11·73	**	13·45	10·54	**	13·67	12·82		13·75
J Individualistic + Group-oriented –	7·26		7·33	9·00	*	7·04	8·64		7·90
O Insecure + Confident –	8·36		9·11	6·82	*	9·39	8·43	*	9·62
Q_2 Self-sufficient + Group-dependent –	8·68		8·16	11·82	***	7·47	9·50	*	7·76
Q_3 Controlled + Uncontrolled –	10·74		9·50	12·64	**	9·51	11·82	**	9·73
Q_4 Tense + Relaxed –	10·79		11·38	10·73		11·44	11·71		11·90
N	42		248	16		164	29		133

Notes

*** Science–arts difference significant beyond the 0·1% level.
** Science–arts difference significant beyond the 1% level.
* Science–arts difference significant beyond the 5% level.

Higher scores are toward the positive pole and lower scores toward the negative pole of each trait. On all scales except B the maximum possible score is twenty and the minimum possible score is zero. Scale B has right and wrong answers with no neutral category, so the maximum score is ten.

The High School Personality Questionnaire (Cattell and Cattell, 1969) measures a set of fourteen dimensions of personality derived from Cattell's 16PF Personality Inventory for use with schoolchildren.[2] It has frequently been used in studies examining the arts–science dichotomy. Form A, which has no time limit for completion, was employed in this investigation. The questionnaire is designed to elicit reactions to carrying out a variety of activities (such as going to a bonfire party and reading a novel). Respondents are asked to indicate whether they like, dislike or are neutral to each activity. There are 142 items (ten for each of the fourteen traits plus two extra items). The items have a neutral response category in order to avoid the weakness inherent in a forced choice between positive and negative only, which could produce a cumulative unwillingness to answer. Pupils were, however, asked to make minimal use of the neutral category.

A comparison between the girls choosing arts or sciences at the third form stage (Table 11.2) shows the science group to be more dominant and tough-minded than the arts group. While the differences between the two groups in excitability, conscientiousness and control do approach significance, they represent little more than a suggestion of the characteristic scientific personality. By the time these girls reach the fifth form we might expect that a substantial proportion of the group regarding sciences as their favourite subjects would have already decided to specialise in science subjects at A level, if not to follow a scientific career. At this pre-O level stage, when science subjects are becoming more rigorous and abstract, we might anticipate an even greater polarisation of the arts and science groups. This would suggest that those with atypical personality traits might have changed their subject preferences and the differences observed in the third form would be more evident and clear-cut. Certainly, as Table 11.2 shows, the changes over this period in the direction of emphasising differences is quite striking. The science group now differ significantly from those choosing arts on nine of the fourteen traits and exhibit a close correspondence to the stereotypical scientific personality. The girls who favour science in the fifth form were more reserved, emotionally stable, dominant, conscientious, tough-minded, individualistic, confident, self-sufficient and controlled in the third form than the girls who favour arts.

It was interesting to discover from an additional analysis (reported in Meredith and Bradley, 1976) that among the boys in the sample the arts and science groups were not significantly different on sociability and self-sufficiency, whereas the girls' groups were. Indeed, self-sufficiency is the trait which distinguished most clearly between the arts- and science-orientated girls in the fifth form. The profile of significant differences between the arts and science groups among the boys in the sample showed no change between the third and fifth forms. At both stages the science-oriented boys were more intelligent, phlegmatic, dominant, tough-minded and controlled than their peers. It was surprising to find that, while being significantly more tough-minded, the science boys were in fact less self-sufficient and controlled than the science girls.

It would appear that in the third form the group of girls favouring science subjects, although relatively small, was somewhat diffuse in terms of personality. By the fifth form, owing to either the increasingly theoretical nature of the discipline or to the pressures of convention, this group has been narrowed down to a very small set of identifiable science specialists. These girls are less person-orientated than their peers, more self-confident and individualistic, which could explain their opting for what is not traditionally a female domain. By this later stage the girls who had not scored so highly on these three factors had changed their allegiance to more conventional subject areas. That the personality profile of the science boys was less well defined may be due to the fact that this much larger group contained boys who would not stay on until A level or become true science specialists. Many of them must be seen simply as following the accepted pattern for their sex in indicating a preference for science subjects. The small group of sixteen science girls, on the other hand, all expected to stay on into the sixth form and to go on to higher education with the intention of specialising in science. It was encouraging to see that this expectation was in all cases fulfilled, at least as far as the sixth form was concerned.

A comparison between the mean personality scores of the arts and science groups of girls in the sixth form indicates the existence of significant differences on the traits of dominance, self-confidence, self-sufficiency and control (Table 11.2). The fact that some of the significance levels are slightly reduced and that maturity and tough-mindedness prove to be of less importance than was evident in the fifth-form analysis can be explained in part by the inclusion in the science group of the girls also opting for biology. At the earlier stages of field work the personality profile of girls expressing a preference for biology was not so closely allied to the scientific personality as that of girls favouring physics or chemistry. As expected, this undoubtedly has had a blurring effect on the differences observed.

This does not, however, provide a complete explanation, as is evidenced by our results in relation to trait A. Whereas most previous research has shown scientists to be reserved and detached (A-) rather than warm-hearted and outgoing (A+), the science girls in this sample did not differ significantly from the arts specialists on this dimension. The suggestion that another explanation is required here derives from the fact that this same pattern of no significant science-arts difference was also observed in the boys' groups, even when biologists were not included in the sixth form science category. The only evidence suggestive of a degree of detachment among both the male and female science specialists was their higher scores on self-sufficiency (Q_2+), but this trait certainly cannot be regarded as precluding the capacity to relate well to other people. It may be that the interest shown by adult scientists in things and ideas rather than people is the result of prolonged exposure to a discipline of this nature and that this is not yet evident in the school-age groups. On the other hand, it may be that a new breed of scientists is emerging who are more outgoing and involved in social relationships than has been the case in the past.

It will be remembered from the previous stages of analysis, when the pupils were in the third and fifth forms, that for the boys there was no evidence of an increased polarisation into science and arts groups along the personality dimensions, as there was in the case of the girls. It was suggested that this was because—unlike the girls, who were by then an apparently highly defined set of science specialists—the group of boys stating a preference for science probably contained many who would not actually go on to specialise in science. This conjecture seems to gain support from the sixth form results. The group of boys who went on to study science A levels could in fact be differentiated from the arts boys on three more of the personality traits than in the previous analyses—sober-mindedness, conscientiousness and self-sufficiency. In many respects this profile is similar to that of the sixth form sample described in Chapter 12 by Smithers and Collings.

Turning our attention again to the science girls, an overall picture emerges of a group characterised by a high measured intelligence combined with a personality profile that provides a clear indication of the reasons why they remained undeterred by social pressures encouraging conformity to their traditional sex roles.

In general terms the results for both boys and girls suggested that patterns of cognitive ability and personality observed as early as the third form contained much information that prefigured academic specialisation three years later. It seemed possible that with this set of variables alone a good prediction of science specialisation could be effected. A statistical technique designed to explore such a relationship in data of the kind we had accumulated is discriminant function analysis. The technique is described in detail by Rulon (1951) and by Tatsouka (1971). It is basically a way of determining the extent and manner in which two or more defined groups of individuals may be differentiated by a set of variables operating together and of subsequently predicting future group membership.[3] A complete description of our results is reported elsewhere (Hutchings, Bradley and Meredith, 1975) but the main findings as they relate to science specialisation in girls will be briefly outlined here.

A discriminant analysis was performed on the scores obtained on both parts of the third form intelligence test. The scores on Part I (verbal-numerical) failed to produce a significant degree of separation between two sixth form groups. The differentiating power of Part II (spatial) was greater: the separation achieved by spatial ability alone between the science and arts girls was significant beyond the 2 per cent level. It was in fact more important in discriminating between the girls' groups than between the boys'. Despite the fact that the science group was not as clear-cut as it might have been, owing to the inclusion of those girls who had chosen to study biology, the degree of separation achieved by the third form personality variables combined in a discriminant function analysis was significant beyond the 1 per cent level. Dominance, control, self-sufficiency and individualism emerged as the main personality variables which could be used in the prediction of science specialisation in girls. The positive side of trait E (dominance) is associated with self-assurance, independence of mind and unconventionality. Consistent with this is the science girls' tendency towards the positive side of traits J and Q_2. Whereas the person whose scores fall on the negative side of these traits tends to value social approval and conventionality, a high positive score represents a confident and resourceful self-sufficiency. Q_2+ individuals are resolute and accustomed to making their own decisions. Finally, high scores on the positive side of trait Q_3 are associated with self-control, forethought and ambition. While sober-mindedness was among the main predictors of science specialisation in the boys' groups, this trait was not an important distinguishing characteristic between the science- and arts-oriented girls. Self-sufficiency, on the other hand, had little discriminatory power among the boys.

The results of this analysis, when considered in conjunction with the longitudinal data on subject preferences, point to the conclusion that active steps must be taken if we wish to encourage girls to regard the sciences as a real alternative to their more traditional options. It would appear that even by the third form the vast majority of girls have already decided that the physical sciences are not for them. As we have seen, the proportion reversing this decision by the fifth form is miniscule. The findings relating to cognitive ability indicated that a much larger proportion of those who had expressed an early interest in science subjects could in fact have gone on to study physics and chemistry at a higher level. Instead they moved to other subject areas or to the more 'socially acceptable' science option of biology. This, then, would suggest that a programme of intervention in the middle years of secondary school to combat the effect of social pressures might encourage more girls to follow up their early interest in the physical sciences. For this group the main emphasis should probably be placed on measures calculated to counteract the masculine gender identity with which science is currently associated. For the majority, however, it is clear that interventionist strategies designed to stimulate an initial interest in science subjects will have to be brought into operation at a much earlier stage. The results presented in Fig. 11.2 showed that for the true science specialists a commitment to science subjects is well established by the third form. Only 17 per cent of the sixth form science group had expressed an early interest in arts or social studies.

Early intervention might take one (or both) of two forms. The present data on personality suggested that, to pursue an interest in science, girls have to be very determined and unhindered by a concern for social approval. The most likely reason for this seems to relate to the masculine image of science both as a career and as a school subject area. If this is indeed the case, then an argument may be made for demonstrating that this image is simply an accident of history rather than intrinsic to the nature of science. There is, however, another side to the current debate which postulates that the roots of girls' rejection of scientific pursuits lie not in social but in biological factors (see Chapter 3). Certainly the data on cognitive abilities indicated that girls generally obtain lower scores on spatial-diagrammatic items. If it is argued that girls therefore lack the abilities necessary to succeed in science, it is within the powers of our educational system to introduce measures to rectify this. The fact that boys do learn verbal skills suggests that initial disadvantages need not persist, yet the insistence on the acquisition of such skills by boys is not matched by a similar form of compensatory education enabling girls to improve their spatial abilities. The whole question of intervention and the variety of forms it might take is discussed in greater detail in the final chapter.

Biased education and training for stereotyped male and female roles is bound to force a significant proportion of both sexes into life styles which are satisfactory neither to themselves nor to society. It must be argued that this represents a major waste of talent. Until full-scale interventionist policies can be implemented the task facing all those responsible for the education and guidance of the young is to challenge the tendency to channel both boys and girls into roles which may seem appropriate for the average member of each sex, but not for the large number of individuals of both sexes whose interests and abilities do not conform to the norm.

Notes

1 See the Statistical Appendix for a discussion of significance testing.

2 Smithers and Collings (Chapter 12) use another derivation from the same Personality Inventory.

3 See the Statistical Appendix for a brief discussion of discriminant analysis.

References

Altmeyer, R. (1966), 'Education in the arts and sciences: divergent paths', doctoral dissertation, Carnegie Institute of Technology.

Cattell, R. B., and Cattell, M. D. L. (1969), *Handbook for the Jr-Sr High School Personality Questionnaire*, Illinois: Institute for Personality and Ability Testing.

Cattell, R. B., and Drevdahl, J. E. (1955), 'A comparison of the personality profile of eminent researchers with that of eminent teachers and administrators and of the general population', *British Journal of Psychology*, 46, pp. 248–61.

Chambers, J. A. (1964), 'Relating personality and biological factors to scientific creativity', *Psychological Monographs*, 68, 7.

Child, D., and Smithers, A. (1971), 'Some cognitive and affective factors in subject choice', *Research in Education*, 5, May, pp. 1–9.

Heim, A. (1970), *Manual for the AH4 Group Test of General Intelligence*, revised edition, Windsor: NFER Publishing Company.

Helson, R. (1971), 'Women mathematicians and the creative personality', *J. Consult. Clin. Psychol.*, 36, pp. 210–20.

Hudson, L. (1966), *Contrary Imaginations: a Psychological Study of the English Schoolboy*, London: Methuen.

–(1968), *Frames of Mind: Ability, Perception and Self-Perception in the Arts and Sciences*, London: Methuen.

Hutchings, D. W., Bradley, J., and Meredith, C. (1975), *Free to choose: Origins and Prediction of Academic Specialisation*, Oxford University Department of Educational Studies.

Maccoby, E. E., and Jacklin, C. N. (1974), *The Psychology of Sex Differences*, Stanford University Press.

Meredith, C., and Bradley, J. (1976), 'A consideration of arts–science personality differences with particular reference to the thing–person dimension', *Educational Studies*, 2, 1, pp. 33–4.

Ormerod, M. B. (1969), 'Personality and ability patterns in arts and science sixth formers', M.Ed. thesis, University of Manchester.

Ounsted, C., and Taylor, D. C. (1972), *Gender Differences: their Ontogeny and Significance*, Edinburgh: Churchill–Livingstone.

Roe, A. (1953), 'A psychological study of eminent psychologists and anthropologists and a comparison with biological and physical scientists', *Psychological Monographs*, 67, 2.

Rossi, A. S. (1965), 'Barriers to the career choice of engineering, medicine or science among American women', in J. A. Mattfeld and C. G. Van Aken (eds.), *Women and the Scientific Professions*, Cambridge, Mass: MIT Press, pp. 51–127.

Rulon, P. J. (1951), 'Distinctions between discriminant and regression analysis and a geometric interpretation of discriminant function', *Harvard Educational Review*, 21, 2.

Saville, P., and Blinkhorn, S. (1976), *Undergraduate Personality by Factored Scales*, Windsor: NFER Publishing Company.

Smithers, A. (1974), 'Science/non-science differences and their relation to academic attainment', in D. E. Billing and J. R. Parsonage (eds.), *Research into Tertiary Science*, London: SRHE, pp. 4–10.

Tatsouka, M. M. (1971), *Multivariate Analysis: Techniques for Educational and Psychological Research*, New York: Wiley.

Thistlethwaite, D. L. (1973), 'Accentuation of differences in values and exposure to different fields of study', *Journal of Educational Psychology*, 65, 3, pp. 279–93.

12

Girls studying science in the sixth form

ALAN SMITHERS and JOHN COLLINGS

In this chapter we turn our attention to the girls, the rather few girls, who do stay on into the sixth form to study the sciences, and we attempt to construct a psychological profile of them. Since these girls, particularly those studying the physical sciences, are doing something unusual for their sex, they may well have distinctive and identifying qualities. And, at the same time, there may be a price to be paid. From the characteristics of these girls it may be possible to learn something of the dynamics of science choice.

Psychologists have used a variety of metaphors in psychological description. Some, like the behaviourists, have been content to regard the person as a more or less complicated mechanism, while others have acknowledged greater human powers. Our study has been guided by Ruddock's (1972) differentiated conception of the person, as modified by Kitwood (1977). This takes account of five aspects of the person which are brought together in the concept of 'identity' or degree of inner completeness. It is a dynamic model which recognises that action involves anticipation of possible future states. This is, in our view, a necessary standpoint for studies of subject and occupational choice.

We can see the relevance of this model if we look at what has been learned of the male scientist. Although this is apparently a well-researched area, most of the studies have been concerned with only one aspect of the person, namely 'personality' (conceived broadly as character, ability and temperament). Scientists tend to have high measured intelligence (Duckworth, 1972; Hutchings *et al.*, 1975), to be convergent in their thinking (Hudson, 1966; Pont, 1970), and to be emotionally stable, tough-minded, self-sufficient and reserved (Saville and Blinkhorn, 1976). Roe (1951a, b) has looked at the characters of eminent scientists and found them to be distinguished by their preference for dealing with ideas and impersonal issues rather than people. She also brings out their masculinity, and willingness to work long hours in pursuit of projects.

In addition, some attention has been given to perceptions of scientists by others. Hudson (1968), for example, found that sixth form schoolboys tended to see figures fom the arts as imaginative, warm, exciting and smooth, and figures from the sciences as dependable, hard-working, valuable and manly. In several ways the profile of the scientist that has emerged—masculine, cold, hard, impersonal—is the antithesis of what it is to be feminine in our culture. (See Weinreich-Haste, Chapter 16 for further discussion of this issue.) One question of interest to us was: do those girls entering science share the characteristics of male scientists or are they different in important respects? Our study was partly based on tests and questionnaires, and partly on in-depth interviews.

Altogether, almost 1,900 lower-sixth-formers in twenty schools in the north of England took part (see Collings, 1978, for details). The study was so designed that girls studying science only (defined initially as any combination of science subjects with maths being included with the sciences) could be compared with girls taking non-science subjects, with girls taking a mixture of science and with boys studying the sciences. The girls in science were not all that easy to find, and we decided to concentrate on schools with fairly large groups of girl scientists. The criterion used was that at least 20 per cent of girl A level entrants in 1974 should be scientists. In this way we arrived at a sample of schools which was not strictly random; nevertheless it gave a fair reflection of those schools where girls had the opportunity to study the sciences and were availing themselves of that opportunity. Although the then Direct Grant schools were over-represented, comprehensive and grammar schools, and sixth form colleges were also included. Where a single-sex girls' school was included, as far as possible the equivalent boys' school was also taken.

The results of the testing showed the girls in science to be more intelligent, to have a distinctive temperament, and to be less person-oriented as compared with the other girls. Intellectual abilities were measured using the AH5 Test of High Grade Intelligence (Heim, 1956, p. 68). The scores of the girls, and boys in science, are shown in Table 12.1. It is clear that, as with the simpler AH4 test used in Bradley's study (Chapter 11), the girls in science obtained, on average, higher intelligence test than the girls taking other subjects (though with the larger samples used here Bradley's finding that the science girls scored particularly highly in the spatial part of the test was not replicated). Girls in science also obtained a higher mean score than the boys in science on the verbal and numerical part of the test, but there was no difference on the diagrammatic part.

Table 12.1 Mean intelligence test scores of sixth-formers

AH5 Test of Intelligence	*Boys in science*	*Girls*		
		Sciences	*Other subjects*	*Mixed*
Verbal + numerical	15·3	17·1	13·7	15·2
Diagrammatic	20·3	20·5	16·7	18·9
Full-scale	35·6	37·6	30·4	34·1
N	327	254	596	260

Note. All differences except diagrammatic scores of boys and girls in science are statistically significant beyond the 5% level. (See the Statistical Appendix for a discussion of significance testing).

When intelligence test scores were combined with the overall score from two fluency tests ('uses of objects' and 'meanings of words'—Getzels and Jackson, 1962) to identify convergers and divergers in the manner described by Hudson (1966), a clear relationship with subject area was found. Although 70 per cent of each group were all-rounders, pupils studying science were much more likely to be convergers than pupils studying other subjects. Among girls in science there were about four convergers for every diverger, and the figure for boys in science was similar. Among girls in other subjects divergers outnumbered convergers by five to one. Earlier studies (Hudson, 1968; Cuthbert, 1973; Anderson and Cropley, 1966) have not obtained such clear-cut results for girls.

Girls in science generally had superior performance in O level examinations to the other girls and also to the boys in science. On average, they had taken more O levels, obtained more passes, obtained a higher overall mean grade, and a higher mean grade in the sciences and mathematics. They reported experiencing less difficulty with O level subjects than other girls and they expressed more interest in the sciences and mathematics. As compared with the boys in science, they actually took, on average, fewer O levels in science and mathematics, though more subjects overall. They expressed more interest in the sciences and mathematics that they

Table 12.2 Comparison of mean personality scores of science sixth-formers and other sixth-formers

Personality traits on 16 PF test	*Girls*	*Boys*	*Summary of previous Studies*
A Outgoing + Reserved –	–	(–)	–
B Intelligent + Unintelligent –	+	+	+
C Emotionally stable + Affected by feelings –	(+)	+	+
E Assertive + Humble –	–	(–)	+
F Happy-go-lucky + Sober –	(–)	–	–
G Conscientious + Expedient –	+	(+)	+
H Venturesome + Shy –	–	–	n.d.
I Tender-minded + Tough-minded –	–	–	–
L Suspicious + Trusting –	–	–	n.d.
M Imaginative + Practical –	–	n.d.	–
N Shrewd + Forthright –	n.d.	(–)	n.d.
O Apprehensive + Self-assured –	(–)	–	–
Q_1 Experimenting + Conservative –	n.d.	n.d.	n.d.
Q_2 Self-sufficient + Group-dependent –	n.d.	(+)	+
Q_3 Controlled + Undisciplined –	+	+	+
Q_4 Tense + Relaxed –	–	–	–

Notes

\+ Scientists score higher (i.e. towards positive pole) than non-scientists. Difference significant beyond the 5% level.

– Scientists score lower (i.e. towards negative pole) than non-scientists. Difference significant beyond the 5% level.

n.d. No difference.

() Differences not reaching 5% level of significance, but greater than 0·15 of a standard deviation.

had taken.

In case it should be that these are more the characteristics of girls in Direct Grant schools than girls in science, the analyses were repeated on just the pupils in co-educational State schools and the same picture emerged: the girls in science tended to have high measured intelligence, to be convergent and to have better O level examination performance.

The temperament of the pupils was studied using Cattell's Personality Inventory (Cattell *et al.*, 1972). Table 12.2 brings out the great similarity in the personality profiles of the boy and girl science choosers as compared with their peers. In both sexes the science choosers were found to be more intelligent (B+), tough-minded (I–), controlled (Q_3+) and relaxed (Q_4–). In addition, the boys in science scored significantly higher than their peers on emotional stability (C+), sobriety (F–) and self-assurance (O–), with the girls in science scoring in the same direction; the girls in science, as did the boys, came out as more reserved (A–) and conscientious (G+). All of which are consistent with previous descriptions of scientists. Both the girls and boys in science came out as less assertive (E–) and more shy (H–) and trusting (L–) than their peers, and in this they tended to differ from the typical science profile.

In order to explore levels of person orientation, in the sense of a preference for dealing with, or involving oneself in, emotional, social and interpersonal issues, as distinct from impersonal ones, a scale was especially devised for this study (Collings, 1978). Table 12.3 shows that, in general, the girls were more person-oriented than boys, and in each case the science group was the least person-oriented. This is consistent with previous findings (Rossi, 1965; Helson, 1971). Roe (1953) has attributed the low person orientation of scientists to early experiences, and makes the point that a higher

Table 12.3 Mean person orientation scores of sixth-formers

	Sciences	*Other subjects*	*Mixed*
Girls	68·2	77·7	72·4
Boys	57·7	65·5	61·4

Note. The differences between girls and boys and between students studying different subject areas are significant beyond the 5% level. The gap between girls' and boys' scores did not vary significantly between subject areas.

proportion than to be expected were only or first-born children. However, no evidence of this for either boys or girls was found in the present study.

The images pupils held of the arts and science sixth-former were ascertained from ratings on five-point bipolar adjectival scales. Table 12.4 shows that there was remarkable correspondence between all groups. Girls and boys in the sciences, other subjects, and the sciences in combination with other subjects all indicated the science sixth-former of their own sex to be more intelligent, hard-working, valuable, and less imaginative than the arts sixth-former of their sex. And, significantly, all groups also saw the science sixth-former (of their own sex) as the more masculine. On the other two

Table 12.4 Comparison of the image of arts and science sixth-formers of their own sex held by sixth-formers studying different subjects

Dimension	*Girls studying*			*Boys studying*		
	Sciences	*Other subjects*	*Mixed*	*Sciences*	*Other subjects*	*Mixed*
Exciting + Dull–	S	A	A	S	A	A
Imaginative + Unimaginative –	A	A	A	(A)	A	A
Hardworking + Lazy –	S	S	S	S	S	S
Intelligent + Unintelligent –	S	S	S	S	S	S
Feminine + Masculine –	A	A	A	A	A	A
Valuable + Worthless –	S	S	S	S	(S)	S
Attractive + Unattractive –	(A)	A	A	S	A	A

Notes

S Scientists rated higher (i.e. towards positive pole) than artists. Difference significant beyond the 5% level.

A Artists rated higher (i.e. towards positive pole) than scientists. Difference significant beyond the 5% level.

( ) Differences below the 5% level of significance.

scales there was some tendency to claim the more desirable characteristic for one's own group. Both girl and boy science choosers saw science sixth-formers as the more exciting, but the other groups saw them as the more dull and the arts sixth-formers as the more exciting. Similarly, boys in science saw science sixth-formers as the more attractive, while for the other groups it was the arts sixth-former. However, the girls in science seemed inclined to agree that arts sixth-formers were the more attractive.

The pupils' perceptions of themselves were also studied using adjectival scales. Table 12.5 shows that girls and boys in science saw themselves as more intelligent and better at mathematics than did their peers studying other subjects. The girls in science also rated themselves as less feminine, less attractive, less popular and less sociable. That is, they appeared to see themselves as less socially attractive than their peers.

Whether this is a price they have to pay for doing something odd and masculine like studying the sciences, or whether the sciences are perhaps acting as a refuge for girls who want to be judged more in terms of their intellectual abilities than their personal qualities, is not clear from the data. Probably something of both is involved. That the sciences can act as a (temporary) haven is illustrated by this comment by one of the girls when asked how she saw herself in ten year's time: 'Ugly, fat, small, old, marching around a parade ground' (she did not intend going on with the sciences after school but was thinking of going in the WRAF) 'but I dream of being, and would like to be, tall, slim, energetic, and working for the BBC in either light or sound, and having a good social life'.

Girls in science appear, in general, to experience more difficulty in everyday situations involving meeting people than other girls, or boys in science. Using a scale adapted from Bryant and Trower (1974), which asks about the degree of difficulty encountered in thirty specific interpersonal situations, it was found that the girls in science obtained, on average, a social difficulty score of 28·4 compared with 24·6 for girls in other subjects and 25·7 for boys in science, the differences being statistically significant. We also asked the girls about their hobbies and interests, and while there was no overall difference in the number of social hobbies mentioned, interestingly, the girls in science tended to prefer those governed by explicit rules—for example, sports—as opposed to more diffuse social activities like going to parties and being with a group of friends.

It might be thought that girls in science would see less distinction

Table 12.5 Comparison of the self-perceptions of science sixth-formers and other sixth-formers

Dimension	*Girls*	*Boys*
Excitable + Calm –	A	A
Intelligent + Unintelligent –	S	S
Sociable + Unsociable –	A	n.s.
Self-confident + Lacking in self-confidence –	n.s.	n.s.
Masculine + Feminine –	S	n.s.
Ambitious + Unambitious –	S	S
Popular + Unpopular –	A	n.s.
Dominant + Submissive –	n.s.	n.s.
Successful + Unsuccessful –	n.s.	S
Good at maths + Poor at maths –	S	S
Strong personality + Weak personality–	n.s.	n.s.
Attractive + Unattractive –	A	n.s.
Like my father + Like my mother –	n.s.	S

Notes

S Scientists rated themselves higher (i.e. towards positive pole) than did non-scientists. Difference significant beyond the 5% level.

A Non-scientists rated themselves higher (i.e. towards positive pole) than did scientists. Difference significant beyond the 5% level.

n.s. Difference not significant at the 5% level.

between the roles of the sexes than other girls, but this does not appear to have been the case. Using a list of thirty interests, activities and occupations (from Hudson, 1968) the sixth-formers were asked to indicate what they regarded as suitable for young men only, suitable for young women only, suitable for both, or for neither. On average, boys indicated that twelve items on the list were suitable for only one of the sexes while the girls, on average, designated nine of the items in this way. There were no differences between pupils studying different subject areas in the way they categorised items. Several of the girls in science indicated that 'being a physicist' or 'being a doctor' was unsuitable for young women before crossing out their first thoughts!

The pupils were asked about their plans for the future. Although they were approached at an early stage in their sixth form courses when post-school education and the world of work may have seemed a long way off, some marked differences between the groups emerged. Nearly all the girls in science (over 80 per cent) were hoping to take a degree course after school, as compared with only about half the other girls. Girls and boys in science were similar in this respect. However, when the preferred field of higher education/employment was examined it became apparent that the girls were mainly aiming for the biological and health sciences and the boys for the physical sciences. Very few of either the boys or girls in science seemed attracted to teaching (2 per cent and 5 per cent respectively).

In giving reasons for their choice of O levels, interest was equally important for all groups of girls, but less important for boys in science than for girls in science. Career was equally important for girls and boys in science, but less important for girls in other subjects.

So far our definition of being 'in science' has been rather broad: taking two or more science subjects at A level and no arts subjects. That is, those specialising in the biological and physical sciences have been taken together. When those taking a biological subject are separated from those taking no biological science, some interesting differences emerge. For example, as Table 12.6 shows, girls in the physical sciences seem to have chosen their subjects more out of interest, and less with a career in mind, than those in the biological sciences. Thus the high career orientation in the plans of girls in science reported above seems mainly to be that of girls intending to go into medicine and the health sciences, and taking biological sciences at school by way of preparation.

Table 12.6 Reasons for A level choice

Reason	*Girls in physical sciences* %	*Girls in biological sciences* %
Interest	50	36
Organisational	6	1
Teachers, family, peers and higher education	13	12
O level results	18	17
Career	13	34
N	126	289

Note. Pupils could give more than one reason for their choice, so the total number of reasons given is greater than the total number of girls studying science. The percentages give the percentage of the total number of reasons which fell in that category, *not* the percentage of girls who gave that reason.

The portrait of girls in science in relation to personality and perspective is also sharpened by considering only those girls in the physical sciences. The girls in physical science had higher intelligence test scores, did better in O level science and mathematics examinations, emerged as more reserved, self-assured and relaxed in temperament, and were less person-oriented. They perceived themselves as less feminine, less sociable, less popular and less like their mothers than the girls in biological sciences.

The definition of 'in science' can be narrowed further by considering only those who intended to continue to study or to work in the sciences on leaving school. Again the profile sharpens, with the 'stayers' having the higher intelligence test scores, obtaining the better mean O level grade in science and mathematics, being more tough-minded, self-sufficient and controlled in temperament, and scoring lower on person orientation. They saw themselves as less feminine and less attractive.

A picture thus emerges of girls studying the physical sciences in the sixth form as:

1. Being highly intelligent and convergent.
2. Having particularly good academic records generally and in the sciences.

3. Being stable, tough-minded, and somewhat introverted.
4. Not being very person-oriented.
5. Perceiving themselves as very able intellectually.
6. De-emphasising their femininity and social attractiveness.
7. Reporting social difficulty.
8. Being relatively little concerned with career considerations.

Girls in science come out, then, as mainly similar to boys in science. In some cases, as with ability and academic record, they are like the boys, only more so. This could be a reflection of the social and psychological barriers they have had to overcome in order to be doing what they are doing. Another instance perhaps of Dale's law (1963) that where a minority has had to prevail against special difficulties in order to enter a level of education, then those who do make it will tend to outperform those of the majority.

Girls in science were also like their male counterparts in being less person-oriented than their peers. Writers like Roe (1959) and McClelland (1962) have suggested that some scientists are not merely uninterested in people, but actually find the complexities of human relationships distressful. In the present study, girls in science (though not the boys) did report experiencing more difficulty in meeting people than did their peers, though whether this was a cause or consequence of subject choice was not explored. The possibility does exist, however, that science choice may involve a 'people avoidance' component.

Where girls and boys in science did differ was in their perceptions of themselves. This was not in relation to abilities, where both groups showed high self-esteem, but with regard to their social attractiveness. Again, it is not clear what led to what, but the girls in science did tend to hold a somewhat negative view of their feminine appeal, and it is possible that the sciences may be acting as a haven.

Do the characteristics of the girls in science give any clues about the possible dynamics of science choice (which, of course, starts long before the sixth form)? Our results may be consistent with a model of adolescent development proposed by Douvan and Adelson (1966) in which identity formation is crucially linked to perceptions of future states. They suggest that the style and focus of future orientation is different for boys and girls: in their view, adolescent boys tend to construct their identities around the work that they will do as men, girls around *who* they will become. They propose that for girls planning is mainly in terms of a developing femininity—making themselves attractive to men and developing general skills relevant to the roles of wife and mother.

Harding argues in Chapter 14 that in deciding to specialise in the physical sciences a girl to some extent must reject this developing femininity principle of integration of personality, and this may lead her to see herself as different, less acceptable to society and to experience greater 'social difficulty'.

This interpretation has several important merits. It neatly and ingeniously brings together many of the characteristics of girls studying physical sciences in the sixth form as reported here. And, notably, it serves to locate subject choice as an aspect of identity formation. But it can only be partial.

In exploring adolescent development mainly in relation to anticipated futures, Douvan and Adelson ignore the effects of present and past experiences—what Ruddock (1972) has referred to as 'personality', 'role' (the parts a person plays in various social settings) and 'perspective' (appraisals based on the reactions of others). Also, although the sharp distinction between the future orientations of boys and girls may have had some relevance to the time and culture to which the Douvan–Adelson study relates (the mid-1950s in the United States), it may be less applicable to present-day England, at least to the very able boys and girls who comprise the present sample. Ruddock's scheme implies that for both boys and girls an identity is chosen on the basis of alternative positions within 'personality', 'role', perceptions of self and others, and 'project' (plans for the future), not just 'project' or 'future orientation' as emphasised by Douvan and Adelson.

Girls in the lower sixth are sixteen to seventeen years old. The shape of their lives will have been constrained to some extent by what has befallen them. They will be looking to their future and developing their self-image, picking up what hints and clues they can from the people and the world around them. The girls in the present study who have opted for the physical sciences are of outstanding ability and have a clear sense of themselves as being very good at the sciences. Their abilities, and their perception of their own abilities, may have been the key elements in the formation of their present identities, overriding any promptings from their social world that the sciences are not really suitable for girls and carrying them forward to a future which they have planned as far as a degree course at university or polytechnic but no further. The best indicator we found in our study of the proportion of girls studying the sciences at any particular school was the mean level of intelligence of girls in that school.

The importance of ability is also suggested by considering the effects of single-sex and co-educational schools on science choice. Several reports (Crowther, 1959; Dale, 1974; DES, 1975) have drawn attention to the fact that proportionately more girls in single-sex school than in mixed schools opt to study the sciences. Several explanations have been offered but a favoured one is in terms of the perceived masculinity of the sciences (see Chapters 7 and 16). It is argued that in a co-educational setting girls become more aware of behaviour thought to be appropriate to their sex and there is more pressure on them to express their femininity. School subjects have somehow acquired gender identities and the physical sciences have come to be seen as masculine. One way girls can demonstrate their femininity is by rejecting masculine preoccupations like studying the sciences.

Our results do show (Table 12.7) that, taken as a group, girls in single-sex schools do more often elect to take the sciences, and sciences combined with other subjects, than girls in co-educational schools. But, as Table 12.8 indicated, this effect is dependent on ability. When the girls were divided into the more able and less able

Table 12.7 The percentage of sixth-formers in single-sex and co-educational schools studying in each subject area

	Subject area			*N*	
School	*Sciences*	*Other subjects*	*Mixed*	*Pupils*	*Schools*
(a) Girls					
Single-sex	24·5	49·9	25·6	718	9
Co-educational	19·9	60·7	19·4	392	8
Total	22·9	53·7	23·4	1110	17
(b) *Boys*					
Single-sex	42·4	34·5	23·1	290	3
Co-educational	41·0	33·6	25·4	497	8
Total	41·6	33·9	24·5	787	11

Note. The difference between the proportion of pupils studying each subject area in single-sex and co-educational schools is significant beyond the 5% level for girls but is not significant for boys.

(by sixth form standards), the interesting result emerged that the school effect does not manifest itself among the more able. Thus, if there is a school effect operating through the expression of femininity of some other means, it appears that the more able girls are less susceptible; or, to put it another way, ability would seem to have an overriding influence.

Nevertheless there does appear to be a loss of girls from science, not only among the less able in co-educational schools but also among the more able irrespective of school type. Fewer of the more able girls than the more able boys specialise in the sciences. Thus it appears that girls with the intellectual ability to study the sciences are not doing so for one reason or another. This can be restated in Ruddock's terms to say that the identity implied by science choice seems to be in some sense incompatible with 'personality', 'perspective' or 'project'. We have seen how it can be incompatible with 'project' (plans for the future), but it is also likely that factors associated with other aspects of the person are important. These could centre around the perceived masculinity and impersonality of the sciences.

If it is thought desirable that more girls should study the sciences

Table 12.8 The percentage of sixth-form girls in single-sex and co-educational schools studying in each subject area, by ability

	Subject area			
School	*Sciences*	*Other subjects*	*Mixed*	*N*
(a) *Most able*				
Single-sex	33·7	40·1	26·2	359
Co-educational	35·7	38·2	26·1	157
Total	34·3	39·5	26·2	516
(b) *Others*				
Single sex	15·3	59·6	25·1	359
Co-educational	9·4	75·7	14·9	235
Total	13·0	66·0	21·0	594

Note. The differences between the proportion of pupils studying each subject area in single-sex and co-educational schools is not significant for the most able girls but is significant beyond the 5% level for the other girls.

for longer, what can be done? This is bound up with the question of to what extent do the sciences have to appeal to a particular kind of person, and of whether their perceived masculinity and impersonality are intrinsic, or largely extrinsic and fortuitious? These are large questions, and we will leave their discussion to other contributors to this volume. Our own view is that the masculinity of the sciences is probably mutable, but their impersonality may well be another matter.[1]

Note

1 The authors gratefully acknowledge receipt of an SSRC grant in support of the work reported here.

References

Anderson, C. C., and Cropley, A. J. (1966), 'Some correlates of originality', *Austr. J. Psychol.*, 18, 218–71.

Bryant, B., and Trower, P. E. (1974). 'Social difficulty in a student sample'. *Br. J. Educ. Psychol.*, 44, 13.–21.

Cattell, R. B., Eber, H. W., and Tatsuoka, M. (1972), *Manual for the 16PF with Bibliographic Supplement*, Champaign, Illinois: IPAT.

Collings, J. A. (1978), 'A psychological study of female science specialists in the sixth form', unpublished Ph.D. thesis, University of Bradford.

Crowther, G. (1959), *Report of Central Advisory Council for Education (England)*, Vol. 1, London: HMSO.

Cuthbert, K. (1973), 'Some differences in the personality, attitudes and cognitive styles of students choosing arts or science specialisms in the sixth form and at university', unpublished M.Sc. thesis, University of Lancaster.

Dale, R. R. (1963), 'Reflections on the influence of social class on student performance at university', *Sociol. Rev. Monogr.*, 7, 131–40.

—(1974), *Mixed or Single-Sex Schools?* Vol. III, London: Routledge & Kegan Paul.

Department of Education and Science (1975), *Curricular Differences for Boys and Girls*, Education Survey 21, London: HMSO.

Douvan, E., and Adelson, J. (1966), *The Adolescent Experience*, New York: Wiley.

Duckworth, D. (1972), 'The choice of science subjects by grammar school pupils', unpublished Ph.D. thesis, University of Lancaster.

Getzels, J. W., and Jackson, P. W. (1962), *Creativity and Intelligence: Explorations with Gifted Students*, New York: Wiley.

Heim, A. W. (1956), *Manual for the Group Test of High Grade Intelligence AH5*, London: NFER (revised edition, 1968).

Helson, R. (1971), 'Women mathematicians and the creative personality', *J. Consult. Clin. Psychol.*, 36, 210–20.

Hudson, L. (1966), *Contrary Imaginations*, London: Methuen.

—(1968), *Frames of Mind*, London: Methuen.

Hutchings, D. W., Bradley, J., and Meredith, C. (1975), *Free to Choose: Origins and Prediction of Academic Specialization*, Oxford University: Department of Educational Studies.

Kitwood, T. M. (1977), 'Values in adolescent life: towards a critical description', unpublished Ph.D. thesis, University of Bradford.

McClelland, D. (1962), 'On the dynamics of creative physical scientists', in L. Hudson (ed.), *The Ecology of Human Intelligence*, Harmondsworth: Penguin, 1970.

Pont, H. B. (1970), 'The arts–science dichotomy', in H. J. Butcher and H. B. Pont (eds.), *Educational Research in Britain 2*, London: University of London Press.

Roe, A. (1951a), 'A psychological study of eminent biologists', *Psychol. Monogr.*, 64, 14.

—(1951b), 'A psychological study of eminent physical scientists', *Genet. Psychol. Monogr.*, 43, 121–239.

—(1953), 'A psychological study of eminent psychologists and anthropologists and a comparison with biological and physical scientists', *Psychol. Monogr.*, 67, 2.

—(1959), 'Personal problems and science', in C. W. Taylor (ed.), *The Third University of Utah Research Conference on the Identification of Creative Scientific Talent*, Salt Lake City: University of Utah Press.

Rossi, A. S. (1965), 'Barriers to the career choice of engineering, medicine or science among American women', in J. A. Mattfeld and C. G. Van Aken (eds.) *Women and the Scientific Professions*, Cambridge,Mass: MIT Press.

Ruddock, R. (1972), 'Conditions of personality identity', in R. Ruddock (ed.), *Six Approaches to the Person*, London: Routledge & Kegan Paul.

Saville, P., and Blinkhorn, S. (1976), *Undergraduate Personality by Factored Scales*, Windsor: NFER.

13

Differential treatment of boy and girl pupils during science lessons

MAURICE GALTON

There are numerous research findings in education which describe changes in pupils' behaviour over a period of time, often from one school year to the next. A large number of these measure attitudes, and many of them are concerned to examine differences in attitudes between boys and girls taking science courses (Kelly, 1961; Wilkinson and Laughton, 1968). Interesting though such findings may be, we can only speculate as to the reasons for such changes in the *absence* of information about the behaviour of pupils and their teachers during the intervening period between the first and second administration of the attitude questionnaire.

The findings presented here, however, concern the events taking place in the classroom and the laboratory during some three hundred science lessons. The data were gathered as part of a large-scale research project designed to examine the effectiveness of different styles of teaching employed during lessons in biology, chemistry and physics (Eggleston *et al.*, 1976). Schools were selected at random, although the selection was subject to certain restrictions. Most modern schools (totalling 13 per cent of our sample compared to the national figure for the year in question of 46 per cent) were excluded, since they did not offer single-subject science courses. More important, because the observers were volunteers from among science method lecturers in universities and colleges of education, it was decided to exclude schools which were beyond a twenty-five-mile radius from the tutor's institution. This meant that city schools predominated and, since secondary reorganisation tended to be less advanced in such areas, the ratio of grammar to comprehensive schools was the reverse of the national figures (see Table 13.1). This and the slightly higher proportion of independent and direct grant schools in our sample accounted for the fact that single-sex schools were also over-represented.

Table 13.1 Percentage of types of secondary school (excluding modern schools) taking part in the research

	School teaching				
Type of school	*Biology*	*Chemistry*	*Physics*	*Total Sample*	*National*[a]
Single-sex	51·4	42·1	58·3	54·1	48·7
Mixed	48·6	57·9	41·7	45·9	51·3
Grammar	47·0	34·6	63·9	50·0	35·9
Comprehensive	34·3	44·8	16·7	30·7	47·5
Independent or direct grant	18·7	20·6	19·4	19·3	16·4

Note

a National figures from *Statistics of Education*, Vol. 4, 1970, Dept of Education, HMSO.

Within each school one teacher was randomly selected from those who agreed to take part. Towards the end of the sampling procedure the selection became somewhat automatic in order to maintain equal numbers of teachers in each of the three subject areas. Although direct comparison with the DES figures proved difficult, because they include all graduate science teachers in modern schools and exclude those from the independent sector, both sets of data show a similar age distribution. The national figures for 1970 gave the mean age of science teachers as thirty-five; half were below the age of thirty-four, with the largest individual group aged twenty-seven. The corresponding figures in this study were thirty-four, thirty-two and twenty-seven respectively. However, the match was not so good when sex differences were taken into account, since our sample contained a higher proportion of women teachers, particularly in biology.

Pupils in their penultimate year of the O level course were given pre- and post-tests of both attainment and attitude in the September and June of the school year. In between, some hundred teachers and their classes were observed once during each term using a specially designed checklist, the Science Teaching Observation Schedule, STOS (Eggleston *et al.*, 1975). The team of observers were trained to a high degree of reliability to record twenty-three different teaching behaviours such as the kinds of *questions* teachers asked, the *statements* they made and the *directives* they gave, together with *initiatives* taken by pupils during these lessons. A cluster analysis isolated three distinct styles of teaching science.[1]

Style I, 'The Problem Solvers', were distinguished by the relatively high frequency of teachers' questions combined with relatively greater use of teachers' statements. This particular pattern of teacher-dominated communication was characterised by the unique features of the group, namely the tendency to ask questions which had to be answered by *constructing hypotheses* and the relatively large number of statements of problems made by these teachers, reinforced by the paucity of pupil-initiated or maintained activities. There was thus comparatively little emphasis upon the informational aspects of science in contrast to problem solving and speculative processes. The picture that emerged was characterised by a style of teaching where the initiative is held by the teacher, who nevertheless challenges the pupils with a comprehensive array of questions, *observational*, *problem-solving* and *speculative*, in both practical and theoretical contexts. The teacher's own statements also reflected this emphasis towards science as a problem-solving activity.

Style II, 'The Informers', were characterised by the relatively infrequent use of questions except those *demanding recall* and the *application of facts and principles to problem solving*. The second of these categories, under the ground rules used in the observation schedule, related to problems with one right answer rather than ones requiring open-ended solutions. There was a relatively high incidence of teacher statements of fact and teacher directions to sources for fact finding. This was mirrored by pupils referring to teachers for the purpose of acquiring and confirming facts, which was one of the unique features of this style. The relatively less frequent use of teachers' statements of experimental procedure combined with lack of questions concerning observation and interpretation of data suggested a non-practical bias for this group of teachers. There was little evidence of pupils consulting each other for any purpose other than that of acquiring information, which tended to support the non-practical fact-acquiring image of Style II teachers.

Style III, 'The Enquirers', had more distinctive features than either of the other two styles. No fewer than five categories were used with greater frequency by Style III teachers, and all these fell within the pupil-directed categories. Pupil-initiated and -maintained behaviour was directed towards *designing experimental procedure*, *inferring*, *formulating* and *testing hypotheses*. The infrequently used categories were all teacher-directed ones, such as questions which were answered either by recalling facts or by applying those facts to solve problems clearly defined in statements by the teacher. The group thus appeared to be 'child-centred' with a vengeance compared to Style I and Style II teachers. Their work was, however, practical and the intellectual level of its content appeared to be high. The description of 'pupil-centred enquiry' might not be an inaccurate reflection of the characteristics of this teaching style. Style III is the one most closely identified with the 'Nuffield approach', which provided the impetus behind the revisions of science curricula during the late 1960s and early 1970s.

Support for these descriptions and the positioning of individual teachers within the classification came from impressionistic accounts of class activities provided by each observer at the end of a lesson. From these accounts some estimate of the time spent by teachers on particular activities was made and expressed as a percentage of the total observation period. Further analysis confirmed that Style II teachers spent a greater percentage of their time on teacher-centred activity than did Style I, who in turn spent more time than teachers in the Style III category. The latter group, however, provided more practical and demonstration activity than Style I, who in turn had more than Style II. These descriptions, simplistic as they were, provided additional support for the validity of the Science Teaching Observation Schedule when used by trained observers under normal classroom conditions.

Having identified the main characteristics of each style, borderline cases (individuals who were at the extreme edges of each cluster) were compared in turn with the central teacher in each style group. They were then allocated to the style where their category totals on each of the twenty-three observed behaviours correlated most highly with that of the central member's profile. When this procedure was complete, approximately half the teachers observed were found to have used Style I but only 19 per cent used Style III (see Table 13.2), despite its association with curriculum reform.

Turning to the tests of attainment and attitude, the former consisted of four sub-tests, two measuring *recall*, one covering aspects of *problem solving* and one dealing with *data manipulation*. Pupils took either the problem solving or the data manipulation sub-test along with one of the recall measures. Thus only half the pupils took any one attainment test, in contrast to the attitude questionnaire, which all pupils completed. Given the relative unpopularity of the physical sciences for girls, a complete differential analysis of attainment between styles and sex was not possible, because of the small numbers of pupils in some of the categories. A further difficulty was the lack of variation in the pupils' scores, because the tests, physics particularly, were found by most pupils to be extremely difficult. For example, the mean on each of the fifteen-item physics tests lay between 3·5 and 4·5. Consequently there was insufficient variation in the scores to share between both styles and sexes. For both sexes together Style I gave the best results on all tests in physics, while in chemistry all the styles appeared to be equally effective. In biology, the most popular science subject with girls, Style III gave the highest mean scores for both sexes on data manipulation and problem solving, but for recall Style I was again superior.

The attitude quesionnaire (Galton and Eggleston, 1973) consisted of five scales. For the purpose of this discussion only the first, measuring general enjoyment of the subject (the fun factor), will be considered in any detail. The maximum score on this scale was 35, and the mean scores lay in a range 23·5 to 25·0, so that there was considerably more variation between individual scores. Pupils were divided into 'above' and 'below' the pre-test mean score, and an analysis of post-test scores using styles and sex as the two main factors was carried out.

In contrast to the attainment tests, none of the attitude scales appeared to favour one particular style overall. The most obvious difference was between those with good initial attitudes (above the mean for the whole sample on the pre-test) and those with poor initial ones. Those with good initial attitudes tended to obtain lower scores on the post-test, while the groups with poor initial attitudes tended to improve their position. As Fig. 13.1 shows, the effect is quite marked, especially for the physical sciences (although this may be partly attributed to the statistical phenomenon of regression towards the mean).

The sharpest differences between teaching styles are shown by the physics pupils with poor initial attitudes. A majority of these pupils might be expected to give up the subject at the end of the O level course, and they are presumably similar in some respects to those pupils who opted not to do physical science. Only Style III accomplishes any significant improvement in the general attitude of these pupils. Although girls with good initial attitudes towards physics appear to be disadvantaged by Style III, there were only two such pupils, and so the result must be regarded as very unreliable. Leaving aside this value, the pattern over the remaining groups in physics is that Style II is the least successful in improving or maintaining the pupils initial attitude, while Style III appears to be

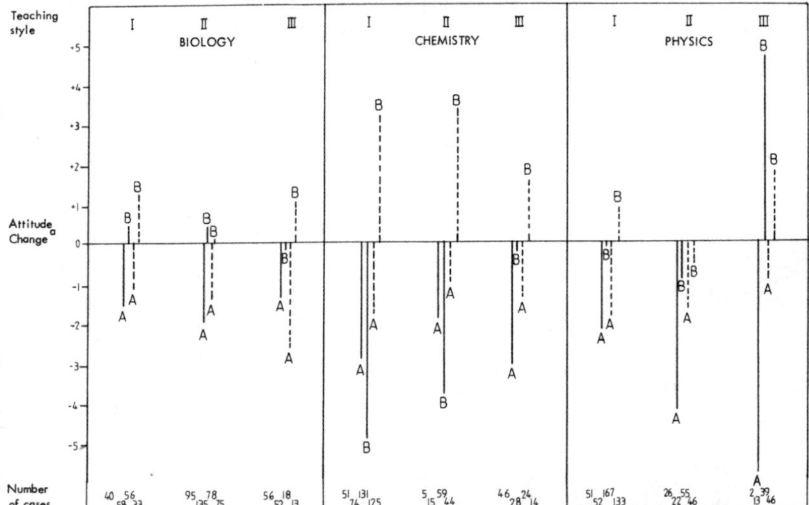

Fig. 13.1 Changes in attitude to science (fun factor) under different teaching styles.

Notes

a The attitude change is the difference between pre- and post-test scores on the fun factor scale (maximum score 35).

A: good initial attitude B: poor initial attitude ____ girls – – – boys.

best. With chemistry the trend operates against Style III and in favour of Style II, except for the girls with poor initial attitudes, where Style III is again superior. Biology shows the smallest changes of all three subjects, with the pupils of both sexes who had good initial attitudes showing a slight fall over the year while the reverse is true of pupils who were below the pre-test mean.

If the results for the attitude scale are combined with those for attainment, then in biology Style III would appear on balance to be the most attractive, particularly in classes where there are both boys and girls. In this sample, however, as we shall see, 52 per cent of the pupils in biology were in fact taught by Style II. In the physical sciences the situation is more complex. In physics the trend is generally in favour of Style III for attitudes but Style I when attainment alone is considered. In chemistry there were no significant differences between styles in attainment, and for pupils with good initial attitudes to the subject it would not therefore seem to matter which style was used. However, for pupils with poor initial attitudes Style III is the best for girls but the worst for boys, so that teachers in mixed-sex classes would appear to have problems.

The question of who uses each style is examined in Table 13.2,

Table 13.2 Distribution of teaching styles by subject, sex of teacher and sex composition of class

	Style I %	*Style II* %	*Style III* %	N
(a) *All classes*	47	34	19	94
Biology	27	52	21	33
Chemistry	58	21	21	33
Physics	57	29	14	28
Men teachers	49	35	16	69
Women teachers	40	32	28	25
Boys' classes	53	33	14	36
Mixed classes	49	31	20	35
Girls' classes	35	39	26	23
(b) *Biology*				
Men teachers	28	61	11	18
Women teachers	27	40	33	15
Boys' classes	33	50	17	6
Mixed classes	29	57	14	14
Girls' classes	23	46	31	13
Men teachers:				
Boys' classes	33	50	17	6
Mixed classes	22	67	11	9
Girls' classes	33	67	0	3
Women teachers:				
Boys' classes	–	–	–	0
Mixed classes	40	40	20	5
Girls' classes	20	40	40	10
(c) *Physical science*				
Men teachers	57	25	18	51
Women teachers	60	20	20	10
Boys' classes	57	30	13	30
Mixed classes	62	14	24	21
Girls' classes	50	30	20	10
Men teachers:				
Boys' classes	55	31	14	29
Mixed classes	65	15	20	20
Girls' classes	–	50	50	2
Women teachers:				
Boys' classes	100	–	–	1
Mixed classes	–	–	100	1
Girls' classes	62	25	13	8

where the distribution of teaching styles by subject, type of school and sex of teacher is presented. The most striking feature of the data concerns the breakdown of teaching style by subject. In the physical sciences Style I predominates, while in biology Style II is used most. One possible explanation for the popularity of biology for girls, in contrast to the physical sciences, is that girls dislike the direct-quesioning problem-solving approach adopted by the majority of teachers of physics and chemistry and prefer either the more open enquiry style which allows them to 'get on with it' among themselves or to be given the facts and told to write them down with a minimum of discussion between the teacher and the class. Support for this view comes from the scores on scale 2 of the attitude questionnaire, which contrasts such positive statements as 'Trying to solve problems in chemistry (biology, physics) is interesting' with a negative view such as 'Chemistry (biology, physics) is just a load of technical terms which are hard to remember'. This was the only scale to show consistent differences between the sexes across all three subjects, with boys having significantly better attitudes than girls to the problem-solving aspects of science. There are a number of studies showing that while liking the teacher appeared to exert little influence on the decision to continue with science (Kelly, 1961; Meyer and Penfold, 1961), when the reverse question was asked there was a strong positive correlation between giving up the subject and dislike of the teacher, presumably indicating dislike of the teaching method also (Pheasant, 1961). The pupils considered here had already made their choice of subjects, but a small-scale study in a middle school showed a similar distribution of teaching styles across subjects (Parks, 1977), so it is possible that teaching styles influence subject choice. The question of why girls dislike the problem-solving approach will be taken up towards the end of this chapter.

Table 13.2 also shows that a higher proportion of men teachers used Style I while a higher proportion of women teachers used Style III. This was partly due to the fact that men teachers were more common in physical science, but there were also marked differences between men and women teachers in biology. Men were more likely to use Style II and less likely to use Style III then women. It has already been shown that, overall, Style III is the most successful for biology, and it has been shown elsewhere (DES, 1976; and see Chapter 14) that girls in single-sex schools reach higher levels of attainment than those in mixed ones. The ratio of women to men science teachers in all-girls schools is generally better than in mixed ones, and a possible explanation for this difference between types of school is that girls in single-sex schools have a greater chance of exposure to the more successful teaching styles.

The distribution of teaching styles between classes of boys and girls is also shown in Table 13.2. Five mixed classes where the ratio of one sex to another was equal to or greater than four to one were classified as single-sex classes. Across all subjects all three types of class have equal chances of being engaged in Style II teaching, but girls are more likely to have Style III, and in both the boys' and the mixed classes Style I predominates. Again this is partly due to the fact that boys' and mixed classes are more common in physical science, but in biology the notion that all-girl classes tend to be taught by the successful style is supported. Although all types of class have a similar chance of receiving Style II teaching in biology, pupils in all-girl classes have twice the chance of being taught by Style III, the most successful biology style.

In the physical sciences there is slightly less chance of girls in single-sex classes getting the most successful Style I. However, this must be set against the previous suggestion that girls may be unhappy with this method of teaching. The numbers of girls' classes and women teachers in physical science are too small for detailed analysis, but in mixed classes the men teachers tend to use Style II which emerges from these results as the least attractive style. Perhaps, as McClelland (1963) argues, however, these highly committed pupils are more like 'human cannonballs and remain undeflected by the vagaries of the teaching'.

Although the evidence is mixed, partly because in some groups the numbers are so small, there is enough here to suggest that the sex of the teacher in combination with the type of class may be a factor in determining the attitudes of girl pupils to science. Men teachers are more likely to use Style I and women Style III. Women teachers tend to teach all-girl classes, so that female pupils in mixed classes are more likely than girls in all-girl classes to have a male teacher who will use Style I.

The fact that Style III gives such varied results with different groups poses problems for the teacher of mixed classes and may be a factor in its relative unpopularity. In biology, using Style III appears to improve the attainment of the girls and boys, but teachers moving away from the more traditional syllabuses, reflecting the Style II approach, may prefer to use Style I as a compromise. In physics there is a different kind of problem in that Style I gives the best results on the attainment tests but, as we have seen, the problem-solving approach does not appeal to girl pupils as much as to boys. In chemistry, however, a decision to use Style I as a compromise, while satisfactory for pupils with good initial attitudes, could lead to problems with girl pupils who are less favourably disposed to the subject. Only the use of Style III appears to raise their enjoyment to anything like a satisfactory level when compared to boys.

Yet in physical science Style I, the problem-solving-directed approach, remains the most popular. There are few documented reports by teachers on the reasons for this but it is likely that the time involved in setting and following up solutions based on pupil-initiated experiments poses severe restrictions, particularly in the context of a two-year course culminating in public examinations. Studies of science teachers in training indicate that the most decisive influence on their teaching style appears to be their own experience as pupils (Eggleston and Dreyfus, 1977). Thus the use of Style I by the majority of teachers may be self-perpetuating from one generation to the next.

The use of a questioning technique, particularly one directed towards obtaining the 'correct answer' as in the problem-solving Style I has potential dangers. There are numerous studies at primary level to show not only that there are differential rates of questioning of pupils in class but that this is dependent on the teacher's perceptions of the pupils' ability (Nash, 1976). In science this has been well documented in a study where the length of time pupils were given to answer a question was recorded (Rowe, 1974). The more able pupils were given longer to answer than the dull ones, presumably because the teacher assumed that the bright pupils were thinking about the answer while the less able ones did not know it and should therefore be spared further embarrassment. However, when Rowe trained student teachers to give more time to these less able pupils they began to answer more questions. It is easy to see how, in the context of the theoretical perspectives developed in earlier chapters, the stereotype of the 'typical scientist' with its masculine connotation could lead teachers of physical science, mostly men, to differentiate their questioning in this way. Unfortunately Rowe did not examine the variation between boys and girls.[2] There is an urgent need for more observational studies of this type which focus on the nature of any differences in treatment afforded to pupils during science lessons. At present, as part of the ORACLE Studies (Galton and Simon, 1975), a number of classes where pupils have recently transferred to secondary school and are experiencing their first taste of 'proper science' are being observed systematically for the effects of such treatment on performance. The possible consequences of this type of 'expectancy effect' should not be in too much doubt. In an earlier series of 'case studies' as part of the development of STOS a teacher, later renamed 'Mr Hisser', was encountered. Pupils in the first science lessons were asked large numbers of questions. Those giving wrong answers would be 'hissed' and 'booed', while the pupils providing correct solutions would be hailed as 'geniuses'. After two lessons pupils who were uncertain about how to answer tended to keep silent when questioned. After a few more lessons the teacher was only likely to call on those he had previously labelled geniuses. The remainder of the class gradually became uninterested and unresponsive. It is not suggested that the majority of science teachers would operate in such an extreme way, but to the extent that research has shown that expectancy effects are likely to be present, to some degree, in all classrooms, girl pupils may be particularly vulnerable under Style I teaching with men teachers. If so, this might explain the popularity of Style III for girls, where the chances of direct intervention by the teacher are minimised and the fact of being encouraged to pursue an investigation independently is, in itself, an acknowledgement by an adult of the pupil's capability.

Notes

1 See the Statistical Appendix for a brief discussion of cluster analysis.

2 But Samuel (Chapter 18) notes that boys are usually quicker than girls in responding to questions, and that teachers may easily come to exclude girls from their lessons by not giving them time or encouragement to reply. [*Editor*].

References

Department of Education and Science (1976) *Curricular Differences for Boys and Girls*, Education Survey 21, HMSO.

Eggleston, J. and Dreyfus, A. (1977) *Teaching Tactics of Student-Teachers of Science*, University of Nottingham, School of Education.

Eggleston, J., Galton, M., and Jones, M. (1975), *A Science Teaching Observation Schedule*, Schools Council Research Series, London: Macmillan.

—(1976), *Processes and Products of Science Teaching*, Schools Council Research Series, London: Macmillan

Galton, M., and Eggleston, J. (1973), *A Modified Science Opinion Poll*, NFER Testing Services, Slough.

Galton, M., and Simon, B. (1975), *Observational Research and Classroom Learning Evaluation*, *ORACLE*, Introductory leaflet, School of Education, University of Leicester.

Kelly, P. J. (1961), 'An investigation of the factors which influence grammar school pupils to prefer scientific subjects', *British Journal of Educational Psychology*, Vol. 31, pp. 43–4.

Meyer, G. R., and Penfold, D. M. E. (1961) 'Factors associated with interest in science', *British Journal of Educational Psychology*, Vol. 31, pp. 33–8.

McClelland, D. C. (1963), 'On the psychodynamics of creative physical scientists', in *Contemporary Approaches to Creative Thinking*, ed. Grubber, M. E., *et al.*, New York: Atherton.

Nash, R. (1976), *Teacher Expectations and Pupil Learning*, Routledge & Kegan Paul.

Parks, D. (1977), 'Styles of science teaching in a middle school', unpublished dissertation, M.A. (Educ.), University of Liverpool.

Pheasant, J. H. (1961), 'The influence of school on the choice of science careers', *British Journal of Educational Psychology*, 31, pp. 38–42.

Rowe, Mary B. (1974), 'Wait time and rewards as instructional variables, their influence on languages, logic and fate control', *Journal of Research in Science Teaching*, 11, pp. 81–94.

Wilkinson, W. J., and Laughton, N. M. (1968), 'Pupils' attitudes to science teaching', *Education in Science*, 26, pp. 31–3.

14

Sex differences in science examinations

JAN HARDING

The curriculum of secondary schools in England and Wales has been influenced by a tradition of subject specialisation by pupils at thirteen-plus and shaped considerably by an examination system that has evolved to provide external assessment for a large proportion of young people at sixteen-plus. In 1976 only 15 per cent of all school leavers had attempted no external examination (before the raising of the school-leaving age, three years earlier, the figure had been 40 per cent). Most schools operate some form of compulsory subjects (which are usually English language and mathematics), supplemented by choices from groups of other subjects. Some large comprehensive schools offer twenty or more options, and, in choosing to study some, pupils inevitably choose not to study others. (The way choices are made has formed the basis of much research, some of which is reported elsewhere in this book (see Chapters 7, 8 and 9.))

The analysis of statistics relating to public examinations at sixteen-plus, therefore, throws some light on the extent to which schools are grounding their pupils in major curriculum components. In this chapter evidence from boys' and girls' entries into public examinations in science, and their performance within them, is discussed and attention is drawn to certain trends that cause concern within the education of girls.

The number of girls and boys entering for sixteen-plus public examinations in science and in other core subjects in 1976 is shown in Table 14.1. For both sexes and in both the CSE and GCE O level the largest number of entries (over 200,000) was in English language. The CSE entries in mathematics approached this figure for both sexes, but the numbers of pupils entering mathematics at O level were lower, especially among girls. Fewer girls entered for O level mathematics than for biology, which may be regarded as a

Table 14.1 Entries by boys and girls in GCE O level and CSE examinations, 1976

	English	*Mathematics*	*Physics*	*Chemistry*	*Biology*	*Technical drawing*
O Level						
Boys	213,248	157,951	109,116	75,820	80,137	48,947
Girls	238,931	112,346	28,813	36,401	129,422	911
CSE						
Boys	238,432	191,588	86,657	43,009	47,309	81,765
Girls	233,092	186,133	11,619	20,418	106,274	1,501

Source. DES, 1977.

girls' subject at this level in that considerably more girls than boys entered for it. Physics, on the other hand, shows a male bias, attracting more entries from boys than any other subject except English and mathematics. The number of girls achieving any sort of grounding in physical science continues to be disturbingly small, especially so for the less able; the ratio of boys to girls entering for physics was four to one and eight to one in O level and CSE respectively (see Chapter 1). Moreover the situation is not improving: the number of boys entering for CSE in physics has increased annually in recent years by more than 2,000, whereas the corresponding figure for girls is only 250; at O level the number of boys' entries has increased five times as much as the number of girls' entries over the last few years.

The Nuffield Science Teaching Project

The Nuffield Science Teaching Project was initiated in 1962 in an attempt to reform school science education. Units were planned for groups of pupils, differentiated by ability and age ranges. The three schemes for physics, chemistry and biology, developed for the O level population, were published four years later. The project recognised the powerful constraining influence of public examinations and negotiated separate special Nuffield examinations in each subject.

In 1971 a research project was set up to enquire into the use of these and other new science curricula.[1] Within this framework I investigated the differential use of the O level schemes for boys and

Table 14.2 National and Nuffield figures, GCE O level, 1972

	Total entries		*Ratio*	*% passes*	
	Boys	*Girls*	*Boys : Girls*	*Boys*	*Girls*
National[a]					
Biology	61,988	106,905	0·6 : 1	61·5	57·3
Chemistry	64,118	28,609	2·2 : 1	61·2	61·9
Physics	92,405	24,420	3·8 : 1	58·6	61·9
Nuffield[b]					
Biology	4,987	4,352	1·1 : 1	70·1	58·2
Chemistry	8,463	3,604	2·4 : 1	69·5	67·6
Physics	9,807	2,304	4·3 : 1	72·1	63·3

Notes
a National figures include Nuffield entries.
b For five Boards only, but representing about 90% of entries.

girls and their comparative performance in examinations for the years 1970–72 (Harding, 1975a).Table 14·2 summarises data collected for 1972 which were representative of a consistent pattern holding in all three years.

It was apparent that the Nuffield courses, especially in biology, were being used more extensively for boys than for girls. The reason for this might lie in the schools and certain characteristics of teachers or in teachers' perception of their pupils' abilities, needs and the role science should play in their education (Harding, 1975b). A considerable number of teachers, when interviewed, commented that girls found greater problems than boys did in using 'the Nuffield approach'. Others denied this and claimed that girls were among their best pupils. The figures obtained for percentage passes appeared to support the former group of teachers, for, although girls did no worse in Nuffield examinations, they did not show the enhanced success in physics and biology that characterised the boys. Perhaps these courses and examinations tapped skills that, for some reason, the girls possessed to a smaller degree, but then the absence of a difference in chemistry was puzzling. Little was known about the nature of the Nuffield sample, except that a large group of entries from boys' public schools was not matched by a similar group of girls.

Support from the Nuffield Foundation, and co-operation from

GCE examination boards were obtained to enquire further into boys' and girls' performance in science, and work began in 1974, using data available on O level science examinations for that year.[2] The remainder of the chapter is concerned with this research and implications arising from it.

Boys' and girls' performance in 1974 O level science examinations.

Six examinations were chosen for further investigation: the three Nuffield O level sciences and one other O level examination in each of biology, chemistry and physics that represented a more conventional course at the time. For reasons relating to the administration of the Nuffield examinations, the 'Nuffield' sample was selected from the candidates entering through one Board only and, for geographical comparability, this Board was identified, in each subject, as that administering the 'conventional' examination.

Within the total populations available, schools were classified by 'intake' (whether single-sex or mixed) and 'selectivity' (whether the school was identified by the DES as grammar, comprehensive or direct grant or independent). Samples of approximately 100 boys and girls from each type of school were drawn systematically for each examination.[3] However, the numbers of direct grant and independent schools entering candidates for Nuffield biology and physics, through the Board chosen for these, were too small to provide viable sub-groups, so these schools were excluded from the main analysis in all examinations.

Table 14.3 shows the percentage pass rate for each sub-group on each examination. No significant *overall* differences between percentage passes for boys and girls were found in any of the six examinations. However, further analyses of the performance of sub-groups showed interesting trends and some significant differences. Most noticable was the variability in the percentage passes of groups of girls. In each examination the range of values obtained for girls was greater than that found for boys. For example, in Nuffield biology only fourteen percentiles separated the best group of boys (those in mixed grammar schools) from the worst group of boys (those in single-sex comprehensive schools), whereas the corresponding figure for girls was thirty-four percentiles. In only two of the examinations did the range for boys exceed twenty percentiles, while that for girls fell below thirty percentiles in only one.

Trends were identified by generating hypotheses and testing these for support or refutation. Differences between sub-groups of less

Table 14.3 Percentage passes for boys and girls by intake and selectivity of school and type of examination in 1974

		Nuffield examination					*Conventional examination*			
		Boys		*Girls*			*Boys*		*Girls*	
Biology										
All schools[a]		58		57			64		64	
Mixed schools		59	>	46	∧		64		61	∧
Single-sex schools		58	<	69			64		66	
Mixed grammar schools		65	>	56	∧	∧	67		67	∧
Single-sex grammar schools		63	<	72			74		73	
Mixed comprehensive schools		54	>	38	∧	∨	60		57	
Single-sex comprehensive schools		51	<	65			55	<	60	
Direct grant and independent schools		–		–			62	<	78	
Chemistry										
All schools[a]		67	>	62			59		59	
Mixed schools		67	>	62			59		61	
Single-sex schools		67	>	62			59		58	
Mixed grammar schools	∧	67		71		∨	66		69	∨
Single-sex grammar schools		73	>	67			59		60	
Mixed comprehensive schools	∨	66	>	54			52		53	
Single-sex comprehensive schools		58	>	50			56		55	
Direct grant and independent schools		61	<	88			78	<	83	
Physics										
All schools[a]		55		54			63		62	
Mixed schools	∧	53		54		∨	66	>	53	∧
Single-sex schools		58	>	54			60	<	66	
Mixed grammar schools	∧	57		61	∧	∨	72	>	60	∧
Single-sex grammar schools		65		68			65	<	75	
Mixed comprehensive schools		50	>	45	∨	∨	60	>	46	∧
Single-sex comprehensive schools		52	>	38			55	<	55	
Direct grant and independent schools		–		–			62	<	79	

Notes

\> Boys do better than girls by at least five percentiles.
< Girls do better than boys by at least five percentiles.
∨ Co-educated pupils do better than single-sex educated by at least five percentiles.
∧ Single-sex educated pupils do better than co-educated by at least five percentiles.
a Excluding direct grant and independent schools.

than five percentiles were discounted, and the remaining differences are indicated in Table 14.3. Tests of significance were applied to these differences for each examination. The hypotheses that obtained support are outlined below.

1. Boys were more successful than girls in mixed schools; the difference was more than five percentiles in Nuffield biology and 'conventional' physics and reached the 5 per cent significance level in the latter examination.
2. Girls in girls' schools were more successful than girls in mixed schools of the same selectivity; the difference was significant for Nuffield biology and for 'conventional' physics.
3. Boys in comprehesive schools were more successful than girls in these schools. This was evident for Nuffield chemistry and Nuffield physics, and, in mixed comprehensive only, for Nuffield biology and conventional physics. However, in these last two examinations and in conventional biology girls did better than boys in single-sex comprehensives. Clearly there is some confounding of the effects of intake and selectivity here.
4. Weakly supported were hypotheses that boys were more successful if they attended a mixed school rather than a boys' school and that girls were more successful than boys if they were educated separately.

Girls and boys performed about equally well in maintained grammar schools.But the girls in the direct grant and independent schools group were outstandingly successful. They gained percentage passes of 80 per cent or more, from relatively large groups of candidates. The same was not true of boys from these schools, who were indistinguishable from groups of boys from other grammar schools.[4]

Both boys and girls obtained higher percentage passes from grammar schools than from comprehensive schools, but this was anticipated and was ascribed to the state of comprehensive reorganisation at the time. As the effect was in the same direction for both sexes, it was not pursued further.

Comparisons of Nuffield and 'conventional' examinations within separate subjects.

Percentage passes for corresponding groups of boys and girls were compared for the Nuffield and more 'conventional' course examinations in each subject. No clear pattern of direction of differences was obtained, and, because the number of schools providing the Nuffield samples was relatively small, any generalisation must be treated with caution. Taking all schools together, both boys and girls gained higher percentage passes in Nuffield biology and chemistry than in the corresponding 'conventional' examinations, but a lower percentage pass in Nuffield physics than in 'conventional' physics. These differences reached significance for both sexes in physics, for boys only in chemistry and for girls from mixed schools only in biology. The pattern shown in Table 14.2 for the 1972 examinations was not replicated in the 1974 data, when comparisons were made between equal numbers of boys and girls from different types of school. There is no hint in these later results that Nuffield examinations are not suitable for girls.

Comparisons within separate parts of each examination

Each of the six examinations used in this study contained more than one part. Although no significant differences between the sexes were apparent in percentage passes, differences *were* found in mean scores obtained by boys and girls in parts of some of the examinations.

The papers included, broadly, three types of item: multiple choice (the two biology examinations contained none of these), structured questions (sections containing questions consisting of several parts were included in this category) and essay-type questions (only 'conventional' biology included questions of this type).

All the physics and chemistry examinations included a separate multiple-choice paper or section, two of them containing as many as seventy items. Mean mark comparisons showed sex differences in favour of boys in half these sections and none in favour of girls.

Structured questions occurred in all six examinations. Four of them showed no difference between the mean marks of boys and girls; in one, boys were more successful only when groups from mixed schools were compared, and in the remaining examination boys were more successful, both overall and in mixed schools.

The 'conventional' biology, paper 2 (the only one to include questions that could be classed as essay-type), was the only part of any examination in which girls were more successful than boys. Thus the relative success of boys and girls in science examinations appears to be influenced by the mode of assessment used: boys tend to achieve higher marks in multiple-choice papers and, while it is not possible to generalise from the one case of essay-type questions, it may be that girls excel in the answering of these. Structured questions appear to show the least bias.

As papers containing multiple-choice items are increasingly being used in assessment, the nature of any sex-related handicap in their operation should be investigated.

Discussion

No support for the overall superiority of performance of boys over girls in science, emerged from this study. However, it must be remembered that these pupils were not a random sample of sixteen-year-olds. Many more boys than girls attempted O level physics and chemistry, so the girls in this group were more highly selected than the boys. The reverse is true for biology, where the boys were more highly selected than the girls.

Conventional wisdom has placed the blame for inadequacies in girls' science education at the door of girls' schools, with their assumed poorer level of provision of laboratories, equipment and staffing. Mrs Renee Short argued in the House of Commons (*The Times*, 1975) that all would be well when mixed schools replaced single-sex schools. But surveys carried out by both Ormerod (1975, and see Chapter 7) and the DES (1975) show that the number of girls choosing to study the physical sciences is a greater proportion of those to whom they are offered when girls are educated alone rather than with boys.

Dale's classic work (1974) reports on studies of the relative success in examinations of boys and girls. All were carried out in grammar schools, and the only studies attempted since the introduction of the General Certificate of Education in 1951 were located in Northern Ireland, where the administration of schools and of the public examination system differ considerably from those in England and Wales.

Dale concluded, '. . . the general pattern established over a period of some forty-five years, with very large numbers of schools and of pupils, and in many and different parts of the country . . shows that the co-educated boys are slightly superior to those boys in boys' schools . . . and girls in girls' schools are approximately equal to the co-ed girls . . .' In several of the cases cited the pupils from girls' schools were significantly more successful than girls from mixed schools, especially in the physical sciences, but Dale emphasises that the girls in the latter schools were handicapped in two of the studies by age, social class and the burden of subjects they studied for examination. He assumes that differentials in these characteristics continue to apply to girls in the different types of school. However,

Rauta and Hunt (1975) found that taking a large number of subjects in public examinations was not necessarily a handicap. In their study, girls taking a large number of subjects passed in a higher proportion of them than did girls of comparable ability taking fewer subjects.

Using data available from the University of London Schools Examinations Board, Wood and Ferguson (1974) have challenged Dale's claims for co-education. The research reported in this paper weakly supports Dale's finding for boys, but strongly reinforces the findings that girls from girls' schools are more successful in science than girls from mixed schools (in spite of the higher level of self-selection, in the physical sciences, of the latter).

Where boys and girls were educated together the boys tended to be more successful than the girls; when they were educated separately there was some support for girls' greater success over the boys. In comprehensive schools there was a tendency for boys to gain higher percentage passes, while there was no evidence of the overall superiority of either sex in grammar schools (Dale's populations). It appears that girls studying science may be at a disadvantage in the mixed school, especially if it is a comprehensive school. Therefore those mixed comprehensives in which girls perform well in the sciences, and do so in relatively large numbers, are of particular interest.

Lavigeur (1976) has drawn attention to two conflicting 'ideologies' of girls' education: one, that they should receive an education equal to that of their brothers; the other, that each sex should be educated to fit the role it assumes in society, which, for the girl, is that of home-maker, wife and mother. The former was expressed in the curriculum of the early grammar schools established for girls in the last quarter of the nineteenth century (of which many of the direct grant girls' schools were linear descendants), while the latter was most succinctly expounded by Newsom (1948, 1963) in the context of secondary modern schools.

When schools are reorganised to become mixed comprehensives the objectives of the teachers and the messages conveyed to the girls become confused. As 70 per cent are 'Newsom children', the home-maker, wife and mother ideology is likely to be pervasive.

In any discussion of girls' achievement in science, two factors must be recognised: that most school subjects (as well as adult occupations) carry a sex bias (which, as Ormerod has shown, becomes more extreme in mixed schools) and that, in spite of cultural claims for science education, the physical sciences have

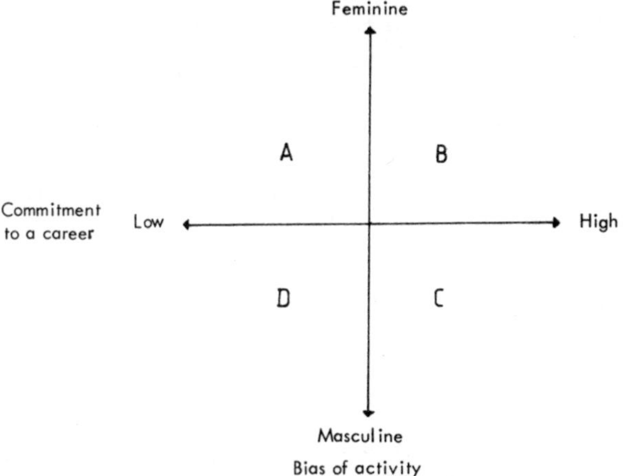

Fig. 14.1 Dimensions relating to the choice of science subjects.

assumed a vocational role in the school curriculum (Harding, 1971).

Bearing these in mind, one can argue that there are at least two dimensions influencing girls' choice of the sciences and possibly their achievement in them. These are *the distribution of school subjects or occupations along a masculine/feminine dimension* and *commitment to a career outside the home*. The first may be constructed from the relative numbers of boys and girls choosing to study each subject or to follow an occupation, but, while the second may be conceptualised, it is more difficult to measure. Its importance, in the argument that follows, is in the increased motivation to succeed that commitment to a career confers. Douvan and Adelson (1966) claim that the boy, in adolescence, uses the job he will do as a principle to make sense of present experience and to integrate his developing personality. For the girl the job merely fills in between school and marriage; the integrating principle she uses is a developing femininity. Such a difference in perspective must surely lead to a differential will to succeed both at school and at work.

Theoretically, the two dimensions defined above may be placed at right angles to define four cells, as shown in Fig. 14.1. Individuals may then be placed within the cells in terms of their position on the career dimensions and their involvement in sex-typed activities.

Cell A will be populated by those who pursue 'feminine' activities with low career expectations, such as secretary, cook, hairdresser or cleaner. Such people will, overwhelmingly, be females.

Cell B will be populated by those who pursue 'feminine' activities, but with higher career expectations. These may be females following careers in teaching, nursing or other paramedical fields, or by males who are at the top of their field in, say, haute-couture, haute-cuisine, music, dance, etc. Cell B is acceptable for females if the career has caring or service characteristics, but acceptable for males only if they assume leadership roles within the activity.

Cell C contains those following careers in 'masculine' activities—largely males. A female in this cell must be prepared to defy convention along both dimensions, and it is this we ask a girl to do when she chooses to study physical science. Such girls may well see themselves as different, awkward and less attractive in the conventional feminine sense. (See Chapter 12.)

Cell D provides no acceptable location for the male, as he would be lacking in ambition. It may contain girls who have chosen to study, say, physics because they like it or are good at it and then are uncertain how to use it, being conscious of the male bias to activities that involve it. Such a girl may decide to teach and therefore move to cell B, as physics teaching is a more acceptable female occupation than practising physics in some other form.

Although that which a girl learns in the home and absorbs in many subtle ways from her contacts with the media and the structure of society may contribute strongly to determining her position on the two dimensions, it is argued that the research reported above suggests that schools are powerful mediators of expectations during the crucial years of adolescence. Their influence is felt through careers advice, the presentation of options and in many unrecognised ways in classroom interaction. The grammar schools, in which Dale and this research detected only small disadvantages for girls, are rapidly being replaced, largely by mixed comprehensives, within which a lesser expectation of girls may prevail and in which, as this research suggests, girls are under-achieving (perhaps not only in science).

The distribution of girls, within a school, in cells A, B, C and D may be a measure of the degree to which a school succeeds in enabling girls to avoid a constraining conformity to stereotypes of female behaviour. Because the physical sciences carry a masculine image, the way girls behave within them may be regarded as an indicator of how that school processes its girls.

Conclusion

The under-achievement of women in science-related careers at all levels, but especially in the more technical fields, has its roots in the school, where few girls obtain relevant qualifications and where many acquire sex-stereotyped attitudes to achievement in work outside the home.

The research into boys' and girls' performance in O level science examinations, described in this chapter, reveals no overall difference in success in terms of percentage passes whatever type of course is followed. However, in maintained schools the most highly selected groups of girls in physical science (those in mixed schools) appear to be the least successful when compared with corresponding groups of boys and other girls. These results suggest that much more research is needed into the ways girls' expectations and performance are shaped in the mixed comprehensive school (the most common school of the future).

There is evidence also that the relative success of boys and girls in examinations may be influenced by the modes of assessment used: multiple-choice items favouring boys and essay-type items favouring girls. We need greater understanding of the skills boys and girls bring to their responses to different types of assessment item.

Notes

1 The Curriculum Diffusion Research Project, supported by the SSRC (1971–74) and based at the Centre for Science Education, Chelsea College, University of London, under the direction of Professor P. J. Kelly.

2 The Girls and Science Education Project, supported by the Nuffield Foundation (1974–75) and based at the Centre for Science Education, Chelsea College, University of London, under the direction of Dr Jan Harding (Harding and Redston, 1978).

3 Systematic sampling means taking every *n*th candidate from a list. The value of *n* is adjusted to yield a sample of the required size (in this case 100). For most purposes systematic sampling is equivalent to random sampling.

4 A large number of boys' public schools enter candidates through the Oxford and Cambridge Joint Examinations Board (not used in this study), whereas few girls' schools do so, and so the samples of DGI boys were possibly deficient in some of the ablest candidates from these schools as a whole.

References

Dale, R. R. (1974), *Mixed or Single-sex Schools*, Vol. III, Routledge & Kegan Paul.

Department of Education and Science (1975), *Curricular Differences for Boys and Girls*, Education Survey 21, HMSO.

Department of Education and Science (1977), *Statistics in Education, 1976*, Vol. 2, HMSO.

Douvan, E., and Adelson, J. (1966), *The Adolescent Experience*, Wiley.

Harding, J. (1971), 'Natural science and teacher education', unpublished M.Ed. dissertation, Chelsea College, University of London.

—(1975a), 'What has Nuffield done for girls?', *Times Ed. Supp.*, 30 November.

—(1975b), 'Communication and support for change in school science education', unpublished Ph.D. thesis, Chelsea College, University of London.

—and Redston, J. (1978), 'Girls and science education report', Library, Centre for Science Education, Chelsea College, Bridges Place, London. SW6.

Lavigeur, J. (1976), 'Educational opportunities for girls, with special reference to co-educational and single-sex schools', unpublished M.Ed. dissertation, Sheffield University.

Newsom, J. (1948), *The Education of Girls*, Faber.

—(1963), *Half our Future*, General Advisory Council for Education (England), HMSO.

Ormerod, M. B. (1975), 'Subject preference and choice in co-educational and single-sex secondary schools', *Brit. J. Educ. Psychol.*, Vol. 45, 257–67.

Rauta, I., and Hunt, A. (1975), *Fifth Form Girls: their Hopes for the Future*, HMSO.

Times, The (1975), 22 January, p. 4, col. 1.

Wood, R., and Ferguson, C. (1974), 'The unproven case for co-education', *Times Ed. Supp.*, 4 October.

15

Girl's science – boys' science revisited

DAVE EBBUTT

In a previous paper (Ebbutt, 1976) partially concerned with attitudes to science in junior forms (aged eleven to thirteen) within the rural comprehensive school at which I teach, I focused my attention predominantly on the mixed ability Form 2 Pembroke of the academic year 1975–76. Within this form I was obtaining disturbing and unexpected glimpses of ideas and attitudes held by the children which were causing them to characterise certain topics within our version of Nuffield Combined Science as either 'girls' science' or 'boys' science'. For instance, crystals and extraction of plant scents appeared to be labelled as 'girls' science'. As the 1977 Christmas term drew to a close I talked to the current 2 Pembroke in an attempt to monitor whether their attitudes to science were causing them to label topics in a similar manner.

I must digress slightly to clarify some basic, but essential, organisational background. This 2 Pembroke class have, in terms of content, followed a similar course to their predecessors of 1975–76 (although teaching personnel differed, in that, as first years, the present form were taught by a young, conscientious male teacher). Thus I had only been teaching them for one term when the discussions, which I will analyse later, took place. Significantly different also was that whereas the former 2 Pembroke covered biological topics in a subject known as environmental studies, which has now disappeared from the timetable, this form have a single period labelled biology. Thus their science time allocation is one double period for physical science plus one single lesson for the biological components of the course. Consequently we could with some accuracy be accused of teaching Uncombined Science even to the extent of using separate notebooks!

This was the form's first taped discussion. Arrangements were rather formal, with the whole class arranged in a ragged half-moon around the front bench upon which I sat with the cassette recorder (Philips N2211) beside me, forgetting somewhat that I was used to the presence of the machine and they were not. I was also conscious that in the 'whole-class discussion' format many of the pupils' replies were liable to be lost by the recorder, and I consciously repeated many of their responses so that the task of transcribing the dialogue would be eased later. The major section of this chapter is concerned with an analysis of this taped discussion, and initially I wish to consider evidence about the physical science components of the course.

Transcript 1

Teacher. OK, I've talked to many classes before, obviously, and what I've found is that they have said—it doesn't mean it's true, but it was true for them—they've said that a lot of science anyway is boys' science.
Chorus of girls. Yer, yeh, yeh. Metals. Battery.
Teacher. If you cast your minds back to last year with Mr P., and perhaps this year, you say metals was boys' science, the car battery was boys' science?
Pupil (unidentified). . . . and the circuits.
Teacher. The circuits. Can you identify—just a minute, Raymond, I'll try and come back to you—but can you identify any areas of science that could be labelled girls' science that you have done?
Several girls. Chemicals. Yes, chemicals.
Teacher. What do you mean by chemicals, Lesley?
Lesley (f). Like looking at copper sulphate crystals, like that.
Teacher. Crystals, right. OK.
Pupil (f.) Tie dye.

The response 'Metals' refers to a topic which was covered during most of the term in which the discussion took place. This is work on the reactivity series of elements and is concerned largely with the physical and chemical properties of metallic elements. The reply 'Battery' refers to an extension of this work into electro-chemistry, completed just prior to the making of the tape, and specifically to an activity of sawing open dry cells followed by a homework to 'research' the structure of a car battery.

Thus the current 2P seem to be delineating various topics as more suitable for girls or boys in a similar manner to, but to a greater degree than, their predecessors. By combining the views of both 2 Pembroke, a perhaps over-simplified arrangement like this emerges:

Activities/topics labelled as girls' science	*Activities/topics labelled as boys' science*
Extracting plant scents (1975–76)	Elements (1977–78)
Making crystals (1975–76 and 1977–78)	Cells and batteries (1977–78)
Tie dye from plant extracts (1977–78)	Electrical circuits (1977–78)

I am aware of the danger of over-simplification in this arrangement and would prefer to think of it as a tendency rather than representing universally held or rigid sexually based views. Indeed, it was not uniformly accepted by all the girls in the form, and I did attempt a very rough head count.

Transcript 2A

Teacher. Is that a universal feeling amongst the girls? Is there any girls that disagree that the science we have done this term (elements) is boys' science?

Pause for five seconds. Three girls who sit together, Karen, Denise and Jackie, indicate their disagreement. Later in the tape I again put the question.

Transcript 2B

Teacher. Is there some extent, then, that some of the girls' opinions got a bit swamped in that big Yeh? How many girls feel that science isn't necessarily boys', and even if it is they don't mind? Right, that's three [the same three]. Three out of about fifteen.

Jumping forward a little in time, during the following term I presented the same class with a hastily prepared questionnaire (Fig. 15.1) during a lull in their normal scheduled physical science practical work. As can be seen, many of the topics or experiments covered by the form over the previous year and a half were listed. They were asked to place the topics in one of four columns—Boys' science; Girls' science; Both; Neither. I then scored the responses to the questionnaire topic by topic. Because this exercise was hastily put together I don't wish to place too much emphasis on the results, so here I have only included those topics which were *unequivocally* seen as boys' science or girls' science, that is, scoring zero in one of those two columns whilst simultaneously a maximum or near maximum score in the other. These results extend the analysis contained in the previous lists.

Fig. 15.1 The questionnaire

I have listened to the tape we made together last term and you seem to me to be saying that certain topics or parts of topics in science or bio are either girls' science or boys' science, for example:

Girls' science	*Boys' science*
Tie dye	Circuits
Crystals	Battery
Plant scents	Elements (metals)
	Dissection

I was wondering about other topics you have done over the past year and a half. Are they boys' science, girls' science, neither or both? Could you write the name of each topic in the appropriate column into which *you* think it fits:

Purifying salt	Weighing substances
Distilling ink	Pressure
Dyes from felt nibs	Spring
Extracting green from grass	Stretching wire
Indicators from plants	Effects of heat on chemicals
Energy–energy changes	Rusting
	Classification of plants and animals
	Oxygen

Girls' science	*Boys' science*	*Both*	*Neither*

Are you a boy or a girl? ____________________

Further topics labelled as girls' science (from the quesionnaire)	*Further topics labelled as boys' science (from the questionnaire)*
Chromatograms from felt-nib pens	Air pressure
	Breaking strain of copper wire
	Energy and energy changes

Turning now to the biological components of the course, the decision to maintain the separate status of biology was a conscious one on my behalf. As I have mentioned earlier, I was sensitised to the boys' science/girls' science issue from the 2P of 1975–76. I had also gleaned evidence from another study in a different school (see Chapter 8) of a greater identification by girls with biology. So I attempted to compensate for some possible alienation on the part of some girls from the physical science components by keeping the biological component separate. The apparent effects of this strategy, combined with the choice of biology topic, is especially interesting and was totally unexpected. I had been roughly following the Combined Science Unit 'Living Things Begin', which includes the structure of flowers, culturing of asexually produced bryophylum plants, an unsuccessful attempt to breed Xenopus toads, through a prepared demonstration of a dissected (male) rabbit, showing the reproductive system. Two rabbits had been brought by the pupils (two girls) on successive weeks and were from the wild. One had been killed by gunshot and was too badly damaged to show—the other was killed by a dog. I had assumed that as the school is in a rural area the pupils would be accustomed to the sight of dead rabbits. By using animals already dead, I would have been able to sidestep any accusation of cruelty on my part. Instead a different discussion followed.

Transcript 3

Teacher. So on the one hand we've got girls, some girls, saying the lads are insensitive; on the other hand we've got the feeling that the girls are too soft. Julie?
Julie. It seems like they're obsessed with death.
Teacher. The boys?
Julie. Yer. [General uproar.]
Teacher. Julie, say that last bit again.
Julie. They all like the blood and gore and everything. They're not interested in life, the way the inside works.
Teacher. So you're saying something very interesting . . .

Lyn. . . . slicing and cutting.
Teacher. Do you hear that, lads? You're not actually interested in what's inside, after all, it's the activity of slicing and cutting which interests you.,
Teacher. Martin . . . no, it's not Martin. Michael, how do you react to that? [Pause.]
Teacher. Milge, how do you react to it?
Milge (m.). I like looking at the guts and that. [Laughter.]

This transcript indicates that the part of the Nuffield Combined Science topic 'How Living Things Begin' which is concerned with dissection can be fairly and accurately placed under boys' science, at least as perceived by 2 Pembroke.

I now want to discuss why some topics should be seen as suitable for boys, and others as more suitable for girls. The pupils' own explanations for this were at a very simplistic level, but they carry with them a clear perception of an acceptable female role/sphere of interest.

Transcript 4A

Julie. My mother didn't do no science, and it hasn't made no difference to her.

or

Transcript 4B

Raymond. Why can't the girls do needlework while we're having science, or something?

This was reflected in a discussion over the worthwhileness of some of the material:

Transcript 5

Teacher. . . . OK. Which of the girls—let's put it really bluntly—which of the girls find the stuff we've been doing in biology is really not important to know?
Boys. Er, look at them, look at them. [About half have their hand up.]
Teacher. How many of the girls feel perhaps it is important to know, even if the method's wrong? Right, about half. And how many of the lads feel that it really is not important to know the stuff I have been trying to teach? [None signify.] Right, can we move on?

My own explanation is rather different. Previously (Ebbutt, 1976) I had written, 'It was apparent that where there was something to show for their work, and where that something could be identified as

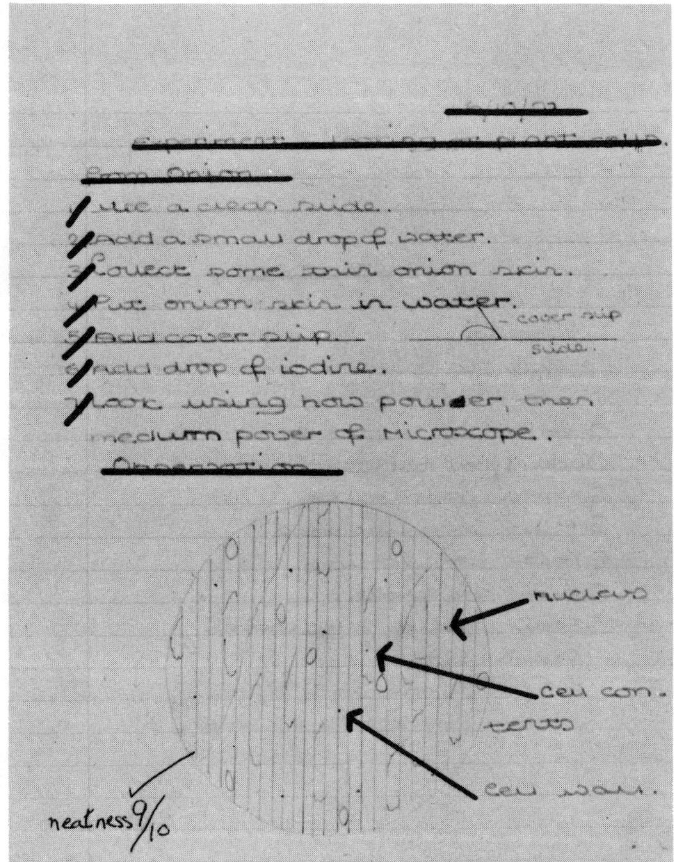

Fig. 15.2 'The neat book syndrome' (B's book).

their own, then they almost universally acclaimed this as enjoyable', and 'Particularly high up on their lists were crystals and dyes extracted from plants'. On closer examination this seems not to be the case in general, but only for girls' science. All the topics labelled by the pupils as girls' science (extracting plant scents, making crystals, tie dye from plant extracts and chromatograms from felt-nib pens) have a tangible end product which fulfills some functional or decorative role.[1] They are nearer to a craft-type activity than the more usual understanding or controlling type of science. I suggest that perhaps this was the underlying criterion which shaped girls'

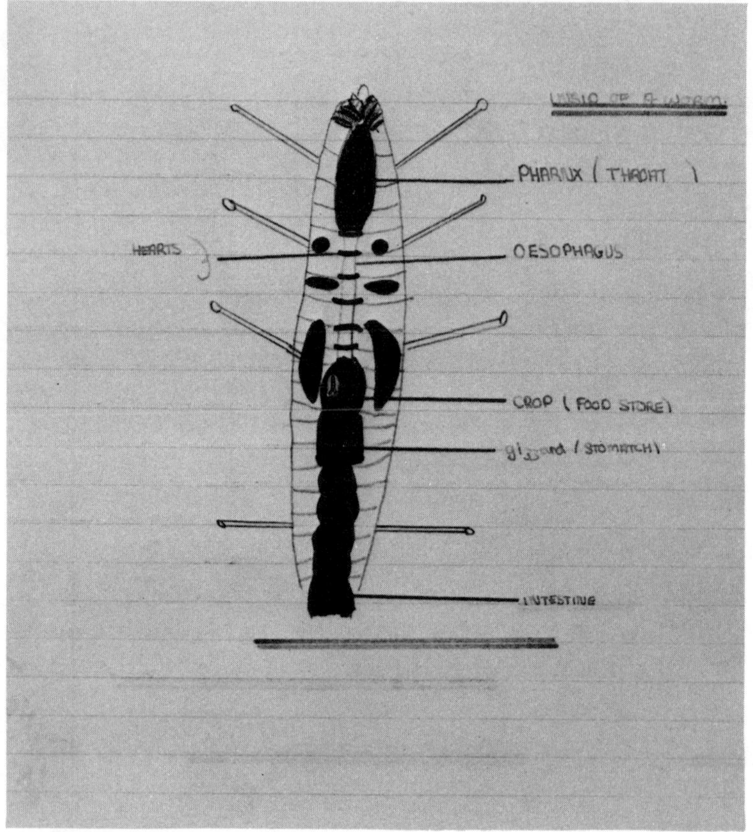

Fig. 15.3 'The neat book syndrome' (S's book).

attitude to science in the junior forms of this school.

Returning briefly to the questionnaire, two topics were seen by the current 2P as unequivocally non-sex-typed, i.e. appearing exclusively in the 'both' column. These were 'Purifying salt', in which the children make their own sample of pure edible salt from 'dirty road salt' and 'Distillation of ink', done as a demonstration, in which pure water is produced and tested. Both of these topics are product-oriented, and so appear to comply with the criterion for girls' science. It is more problematic to speculate upon why they should also be seen as equally boys' science.

If science, in the junior part of the school, has to be product-oriented in an attractive, decorative sense in order to be espoused as girls' science, this might have important implications at a later stage. One possible manifestation is 'the neat book syndrome'. By this I mean the ability which many girls display of producing incredibly neat notebooks (see Figs. 15.2 and 15.3)[2] at the same time demonstrating a depressingly low level of involvement in practical work, or a similarly low level of concept development, or both. The following transcript is taken from a taped 'chat' between myself and two third-year girls:

Transcript 6

Teacher. Now I still cannot get at why it is nicer to produce a nice book than to do a good experiment?
S. Because people look at the books, and experiments you cannot see.
B. You send the books home, don't you? And other people and parents see, and that; and you get your marks. Your marks go on your report from the book, don't they?
Teacher. Now supposing you took that book home to your mum, and showed it to her, and said, 'Look, I've got an A–', now what worries me is that she is going to think that you are a whiz-kid scientist, when in fact what you are is a whiz-kid at is producing neat notes.
B. Oh, yeh. I know what you mean, I know what you mean. She isn't. She just does the writing, copying, doesn't she?
Teacher. She's no scientist, is she?
B. No.
S. No, neither of us, no.
Teacher. And yet, looking through your book, you could get the impression that you were, couldn't you?
B. and S. Yer.
Teacher. Now are you purposely trying to create that impression?
B. and S. No.
S. It's like in physics. I don't like watching experiments, it's boring.
Teacher. But what does your book look like?
S. It looks neater. I don't like doing the experiments.

It seems to me possible to speculate that what is happening in this instance is that the girls' notebooks are becoming *substitute* attractive, decorative end products *in themselves*, in response to the increasing absence of tangible end products as science becomes more complex as the girls progress up the school. Further, it might well be that school biology—at least as it has been traditionally taught in terms of heavy reliance on neat drawing skills—is the science which accommodates this substitution most effectively and is consequently the most popular science option for girls. Extending this idea just a little further, the corollary, within a non-product-oriented science,

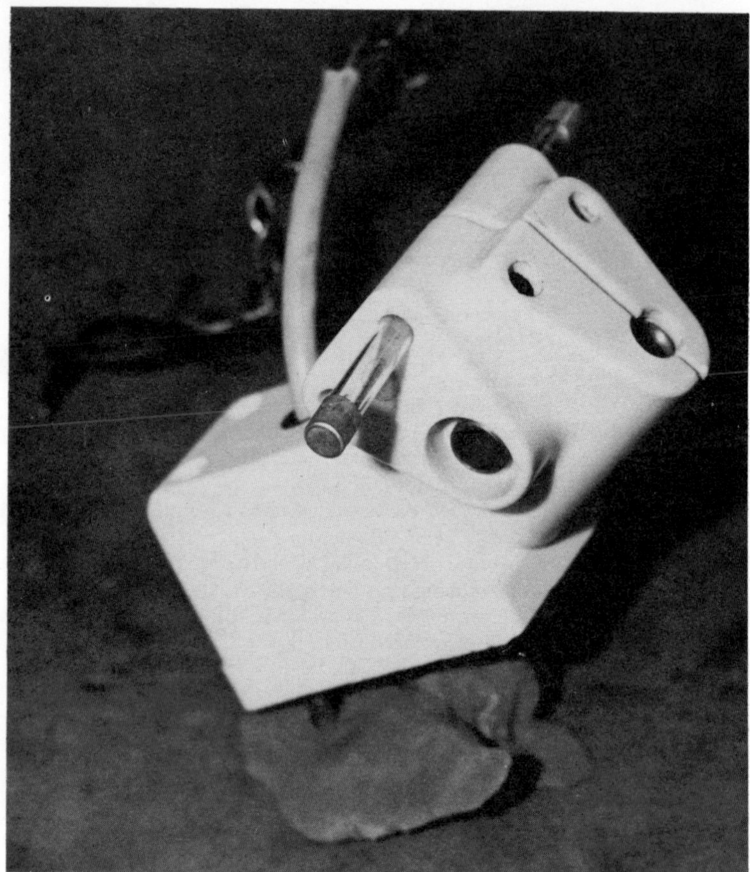

Fig. 15.4 The 'Physics' dog.

such as physics, is that within this school the switch-off by girls, already signalled in the first two years, is complete half-way through the third year. Complete, that is, except for the odd, rare and pathetic attempt to make the subject fit their criteria, as witness a little dog (Fig. 15.4) made from two electrical adaptor sockets, a length of flex and two fuses. This was confiscated by the physics teacher from a group of third-year girls. They had made it during a lesson on the physics behind the correct techniques for wiring plugs on to electrical apparatus such as soldering irons and power packs. The pathos of this particular incident lies in the fact that the manufacture was labelled as deviant.

These observations are suggestive, but not conclusive. They give rise to many further questions, and I will finish with a selection of these questions which future researchers may wish to pursue.

1. Are any of the foregoing observations or conclusions generalisable elsewhere, or are they confined to the micro-worlds of 2P and the third year in this school?
2. If these observations are generalisable, what are the implications for the construction of future science curricula, i.e. separate or different forms of science for girls and boys?
3. Are the attitudes described crystalised by male science teachers?
4. What are the criteria behind the labelling of certain topics as boys' science?
5. Where do the criteria for girls' science come from?[3]

Notes

1 The chromatograms produced by felt-nib pens tend to be concentric rings of delicate, attractive pastel colours, which when formed on strips or discs of filter paper fit neatly and conveniently into pupils' exercise books.

2 A striking feature of these notebook pages was the extensive and varied use of coloured pens. Unfortunately the full effect cannot be conveyed by a monochrome plate. [*Editor*].

3 I would like to thank all the children quoted for their kindness in allowing me to include their contributions; and the Head for allowing this article to be published.

Reference

Ebbutt, D. P. (1976), 'Some aspects of mixed ability science teaching with 2P', *Cambridge J. Educ.*, Vol. 6, Nos. 1 and 2.

16

The image of science

HELEN WEINREICH-HASTE

'Facts and figures in quantity do not interest me. I do not think I would like to be a human computer, programmed to remember facts. Skill and feeling interest me more.' 'I like weird experiments and things going fizzy.' These statements were written by fourteen-year-olds in 1978. They are interesting because, although one is positive and one negative, each reflects an inaccurate and limited picture of the activities of science and the scientist. Yet these adolescents have recently made the choice, which may be central to their future careers, between 'science' and 'arts' options. Other chapters in this book have examined a number of variables which may contribute to girls' under-representation in science; here I shall consider the image of science, and the possible effect it may have on girls' choices and their motivation to do well in science.

The man in the white coat

In a highly technological society, science and technology are very salient. However, our beliefs about science relate mainly to what we think the role of science *should* be; there is a gap between what is believed to go on in science and what actually goes on. This gap is important because it interferes with an understanding of how science progresses. It is also important because it may contribute to the image the growing individual has of the life of the scientist, which may effect his or her choice of career.

From several sources there emerges what might be termed a *consensual cultural image* of science. Science is seen as a *rational* activity. The scientist is objective, and progresses to his conclusions by the careful application of logic and reason.[1] The aims of science are instrumental, to gain control of the environment. Knowledge is an end in itself, but it is also a means to the end of control. In this image, feelings and emotions play little part, and indeed the human and social components tend to be underplayed. Different writers evaluate the image differently, but they tend to agree on its form. (Cotgrove, 1974; Leiss, 1972; Marcuse, 1964; Merton, 1967; Rose and Rose, 1969; 1976).

In this chapter I am concerned with the image of science and the scientist as currently held by schoolchildren and by students. Several studies have indicated that the image of the scientist has two forms. First there is what might be termed the 'stage scientist', a physically unattractive person, usually described as balding, scruffy, old, eccentric and surrounded by laboratory paraphenalia and mess—the 'mad professor' stereotype. The second aspect is, by implication, the kind of person the schoolchild or his contemporaries may grow into, a more moderate and credible image. The latter image, however, still includes lack of social contact, preoccupation with work, and a laboratory milieu. Mead and Métraux (1957) elicited images of the scientist and evaluation of these images, from 35,000 American schoolchildren. The positive image was of a man who was dedicated, careful, involved and highly intelligent, whose labours would be eventually rewarded by personal success and benefit to mankind. The negative image was of a man engaged in boring, repetitive and unrewarding work, out of touch with people and human relationships, and whose intellect dominated all aspects of his life. Even in the biological sciences, the emphasis was upon dead specimens rather than on living creatures.

Hudson (1968, 1970) found that British schoolboys also regard scientists as dull in contrast to the arts man (*sic*). The boys distinguished between value and attractiveness: the scientist was useful to society, worthy, masculine, dull and sexless; the arts man was rather useless, feminine, exciting and sexy. The contrast is not unlike Sellar and Yeatman's (1930) distinction between the Cavalier and the Roundhead; Wrong but Wromantic *v*. Right but Repulsive.

Studies with undergraduates picked up another variable strongly: the *abnormality* of the scientist's image. Research scientists, especially physicists, were perceived as the most intelligent, the most socially isolated and the least 'normal' of scientists. Biologists were closer to the average American male. Engineers were perceived as normal, not particularly intelligent, but also as lacking in social graces (Bendig and Hountras, 1958; Beardslee and O'Dowd, 1961).

The making of the young scientist

The scientist who emerges from these studies is not a very attractive creature; as a career role model, he may be a deterrent to both sexes. It seems quite clear from these studies that to become a scientist is to become a *kind* of person, to take on a whole life style, not merely to take up a job; the image of the scientist is a picture of the whole person, not only of his professional activities. It is therefore a reasonable—if as yet untested—assumption that the process of becoming a scientist means socialisation into the role, not simply training to acquire a meal ticket. For the growing boy, the choice involves his self-image and personality. Studies by Roe (1951) and Eiduson (1962) on the childhood of successful scientists illustrate the possible relationships between these variables. Many of their subjects reported an isolated childhood, with much time spent alone, and social awkwardness with other children. One interpretation of this is that such conditions are particularly conducive to the development of scientific skills and interests. However, it is also the case that such a child may perceive the *stereotypical* image of the life of a scientist as one to which he is particularly suited on the basis of his personality.

For the growing girl, however, there are more complex issues. The world of science is clearly a male world, involving masculine activity. Statistically, scientists are predominantly men, but in addition to this, the image of science and the activities of the scientist are perceived as stereotypically masculine, concerned with military and space applications, and with large-scale engines. Tyler (1964) found that potential scientists were more 'masculine' in their interests than other boys. Furthermore, the constellation of values and attributes associated with science is masculine. In Hudson's study this was explicit; the science man was seen as masculine, and the arts man as more feminine.[2] In a study by Mitroff, scientists involved in the American space programme perceived the stereotype of the scientist (the 'hypothetical scientist') as even more masculine than the 'actual scientist'. The 'hypothetical scientist', to a greater extent than the 'actual scientist', possessed the qualities associated with the cultural image of masculinity—objectivity, independence, logic, ambition, unawareness of the feelings of others, aggression (Mitroff *et al.*, 1977). Schoolchildren hold a cruder stereotype, but with the same components (see below).

When we consider the possible effect of this image on girls' choices about science, it is important that we do not confuse the image of masculinity with the reality. On most of the characteristics mentioned there are no consistent or fundamental differences between boys and girls (with the possible exception of aggression) (Maccoby and Jacklin, 1976). Furthermore, in the education system both sexes are *formally* expected to develop the skills of objectivity, logical ability, independence of thought, and to be ambitious at least within the context of examination success. Therefore the question is not about innate sex differences, but about the extent to which the image of science coincides with

1. The kinds of interests and orientations which have developed by the time the individual comes to make the choice between subjects, and
2. The kinds of expectations which the individual has concerning appropriate and desirable characteristics and activities for the self in adult life.

These are related; extensive research demonstrates a marked difference of interest and orientation between boys and girls, in line with their socialised sex roles. In the case of boys, but not in the case of girls, sex role expectations are congruent with the *image* of science.

It seems from the accumulated evidence (Hoffman, 1972; Weinreich-Haste, 1978 a, b; Chetwynd and Hartnett, 1978) that, for girls, there is conflict between an overt curriculum of achievement and a hidden curriculum which presents 'appropriate' characteristics and qualities for womanhood. This is likely to proscribe a role that is explicitly masculine in its connotations. Other aspects of the image of science have different implications for boys and girls. The exclusive preoccupation with work which is supposed to be characteristic of scientists is more likely, in our culture, to be seen as precluding marriage and a family for girls than for boys, for example.

To resist or overcome the stereotype, therefore, girls need strong motivation, and perhaps other characteristics. There have been a variety of studies with women undergraduates and with professional women, including scientists, on what has been called 'role innovation' in general, and science in particular. A pattern of characteristics emerges from these studies: women who enter non-traditional fields tend to possess high academic ability, independence of mind, reliance on internal rather than external criteria for judgement of self, non-traditional attitudes to sex roles, and to have had mothers who provided a role model of a working woman, or at least of a woman whose orientations were not confined to the role of wife and mother (Hoffman, 1972; Tangri, 1972).

Studies of the science stereotypes

In order to examine the current image of science among British youngsters I undertook two small studies with university students and with schoolchildren around the age at which they make the crucial choice between arts and science. The samples consisted of 260 schoolchildren and 150 university students. In both groups there were equal numbers of males and females. The university students were drawn half from science and technology, and half from social science. The schoolchildren were drawn from two single-sex comprehensive schools. They were aged thirteen and fourteen. The younger children had just made their choice of options for the public examinations; the older children had made theirs a year previously. On the basis of their choices they were classified as predominantly arts specialists or predominantly science specialists. The sample was the whole top band of academic ability in each school in the relevant years.[3]

In the first study the format was the same for both students and schoolchildren, though the content necessarily changed to reflect differences in conceptualisation, and familiarity with different forms of knowledge. The respondents were asked to rate subjects in the arts, sciences and social sciences on a series of six-point scales. The mean ratings for each subject on each scale used with the schoolchildren are shown in Table 16.1. One of the most interesting dimensions is the masculine–feminine scale. There is a clear hierarchy of subjects on this scale, as indicated in Fig. 16.1. Not surprisingly, cooking, typing and woodwork were the most strongly sex-typed subjects, but it is noticeable that physics, maths and chemistry were the most masculine academic subjects. Moreover boys saw all the academic subjects as more masculine than did girls (conversely, girls saw them as more feminine than did boys). Thus each sex appeared to be laying claim to academic subjects. Where comparisons were possible there was quite close agreement between the students' and the schoolchildren's ratings, although the students had a tendency to give slightly more feminine ratings (perhaps surprisingly, in view of the increasing male dominance of all subjects at university).

When correlations between ratings on the various scales were computed, it became clear that there was a cluster of attributes attached to science (Table 16.2). Subjects rated as scientific were

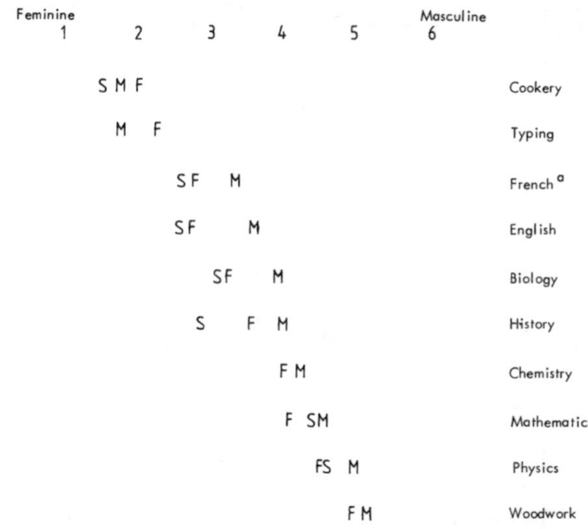

Fig. 16.1 Mean ratings of each subject by schoolboys and schoolgirls on masculine - feminine dimension (with comparison with student sample where applicable).

Notes
F = schoolgirls M = schoolboys S = students
[a] Modern Languages for Students.

also perceived as masculine, hard, complex, based on thinking rather than feeling, and, in the case of the schoolchildren, about things rather than about people. Among the university students the image was consistent: there were no differences in the significance of the correlations for males and females, or for science and social science students.[4] Among the schoolchildren there were, however, differences between the boys' and girls' perceptions, although the general pattern was similar. Girls saw science as difficult, and they also saw complicated and difficult things as masculine. The correlation between science and masculinity was significant for boys but not for girls, but this was mainly because of the anomalous position of cookery (which was rated far more scientific by girls than by boys, although both sexes rated it as strongly feminine). As Fig. 16.1 shows, schoolgirls did in fact see the physical sciences as masculine. Boys, more than girls, found difficult things interesting, and regarded feminine things as boring. Girls more than boys regarded matters concerning feelings and people as simple and easy. In summary, it appears that the sciences are not just rated as

Table 16.1 The mean scores awarded by schoolgirls and boys to each subject on each dimension

		Biology	Chemistry	Cookery	English	French	History	Maths	Physics	Typing	Woodwork	Mean
Interesting–boring	F	1·8	2·7	2·5	2·5	2·9	2·6	2·8	3·4	3·1	2·2	2·7
	M	2·2	3·0	3·4	2·7	3·9	2·7	2·3	2·8	4·4	3·6	3·1
About facts–about ideas	F	1·6	2·2	3·4	4·5	2·4	1·6	1·9	1·8	2·5	4·9	2·7
	M	2·0	2·3	3·9	4·5	2·4	1·9	1·9	2·2	2·9	3·8	2·8
Complicated–Simple	F	2·8	2·2	4·2	3·9	2·9	3·5	2·1	2·0	3·7	3·5	3·1
	M	3·3	2·2	4·3	3·8	2·3	3·2	2·6	2·2	3·4	3·4	3·1
Involves feeling–Involves thought	F	4·7	5·4	4·2	3·3	4·8	4·3	5·7	5·5	4·8	4·1	4·7
	M	4·9	5·6	3·9	3·7	4·7	4·0	5·7	5·6	4·8	4·5	4·7
Difficult–Easy	F	3·2	2·3	4·2	4·1	3·2	3·9	2·2	2·3	3·8	3·5	3·3
	M	3·7	2·7	4·2	4·0	2·4	3·5	3·1	2·6	2·7	3·3	3·2
Science–Arts	F	1·4	1·1	3·4	5·0	4·5	4·4	2·2	1·2	4·0	5·1	3·2
	M	1·4	1·2	4·8	4·8	4·6	4·2	2·3	1·2	4·4	5·0	3·4
About people–About things	F	3·7	5·4	4·2	2·9	3·3	2·0	5·5	5·4	3·8	4·8	4·1
	M	4·2	5·4	4·7	3·4	3·3	2·3	5·3	5·3	4·2	4·9	4·3
Useless–Useful	F	5·3	4·8	5·2	5·4	5·1	4·2	5·6	4·9	5·2	4·8	5·0
	M	4·9	4·7	4·9	5·4	4·6	4·1	5·6	5·5	4·5	4·8	4·9
Feminine–Masculine	F	3·3	4·2	2·1	2·8	2·8	3·6	4·2	4·6	2·3	5·1	3·5
	M	4·0	4·4	1·8	3·7	3·3	4·1	4·6	5·0	1·8	5·4	3·8
Hard–Soft	F	3·5	2·3	4·6	4·2	3·7	3·6	2·1	2·1	3·6	3·4	3·3
	M	3·5	2·5	5·0	3·9	3·5	3·3	2·4	2·2	4·1	2·8	3·3

Note. A rating of 1 indicates complete agreement with the left hand adjective in a pair, while a rating of 6 indicates complete agreement with the right hand adjective. A neutral rating is 3·5.

Table 16.2 The correlations between ratings on different dimensions for school girls and boys

	Interesting–Boring	About facts–About ideas	Complicated–Simple	Involves feeling–Involves thought	Difficult–Easy	Science–Arts	About people–About things	Useless–Useful	Feminine–Masculine	Hard–Soft
							Boys			
Interesting–Boring	–	0·03	–0·13	–0·02	–0·30*	0·20	0·01	–0·39*	–0·27*	0·11
About facts–About ideas	–0·04	–	0·20	–0·29*	0·13	0·42*	–0·08	0·06	–0·14	0·18
Complicated–Simple	–0·14	0·31*	–	0·23	0·69*	0·27*	–0·09	0·00	–0·21	0·34*
Involves feeling–Involves thought	0·08	–0·34*	–0·31*	–	–0·18	–0·41*	0·34*	0·10	0·23*	–0·26*
Difficult–Easy	0·20	0·23	0·74*	–0·29*	–	0·12	–0·08	0·09	–0·04	–0·32*
Science–Arts	–0·04	0·42*	0·40*	–0·38*	0·35*	–	–0·33*	–0·11	–0·33*	0·36*
About people–About things	0·14	–0·05	–0·25*	0·41*	–0·28*	–0·38*	–	0·11	0·15	–0·18
Useless–Useful	–0·43*	–0·03	–0·02	0·11	0·12	–0·02	0·05	–	0·17	–0·14
Feminine–Masculine	0·08	0·00	–0·29*	0·14	–0·29*	–0·22	0·33*	0·12	–	–0·53*
Hard–Soft	–0·17	0·26*	0·56*	–0·28*	0·59*	0·39*	–0·31*	0·02	–0·41*	–
	Girls									

Notes

Boys above the diagonal, girls below the diagonal.

The correlations are computed between individuals, using the BMD program for the analysis of repertory grids.

* Significant beyond the 5% level.

masculine on a scale, but also have other connotations associated with the cultural stereotype of masculinity.

The second study was performed with the schoolchildren only, and followed a similar format to the Mead and Métraux study. The schoolchildren were asked to complete five sentences: 'When I think of a scientist I think of . . .'; 'I would like to be a scientist because . . .'; 'I would not like to be a scientist because . . .; 'I would like to marry a scientist because . . .'; 'I would not like to marry a scientist because . . .'. They were encouraged to complete all the sentences, irrespective of whether they wished to become scientists or not.

The first sentence elicited the most stereotyped imagery. The caricature of a scientist remains virtually unchanged over the past twenty years, apparently. The most frequently occurring elements of this image were *intelligence*, *discovery* or *invention*, *laboratories* and *personal description*. The personal descriptions were very similar to those evoked by Mead and Métraux; 'a mad professor wearing a white coat . . . ginger hair, half bald, talking with a German acent and peering over the top of his glasses', 'a man in a white coat with a bald head and glasses, writing on a clipboard standing in front of a bench covered with apparatus'. Helping mankind, being dedicated and/or introverted were mentioned considerably less frequently. The maleness of the stereotypical scientist is apparent from the fact that over one third of the sample specifically referred to the scientist as a man.

However, when answering why they themselves would like to be scientists, the schoolchildren presented a broader and more positive image. The primary reason given was the interesting nature of the knowledge with which they would deal, and the job itself. Discovery and innovation were also important. Helping mankind was of considerably less importance than the rewards (money or fame) or the joys of experimentation. 'It is an interesting job and exciting when you discover.' 'It would be a great achievement to discover about something that had not been thought of before.' 'People look up to you and I'd feel needed.'

These are eminently reasonable grounds for wishing to be a scientist. The responses reflect a moderated consideration of job satisfactions, with relatively little glamorisation. The negative side ('I don't want to be a scientist because . . .') showed more signs of stereotype and exaggeration. The personal aspects still predominated, in this case the boredom of the job and the subject. However, a close second was the difficulty and complexity of science; again the assumption that only vastly superior intellects need apply. In contrast with the stereotype of science facilitating control over nature, many expressed hopelessness at the possibility of any real advance. 'Science is very unfeeling and the physics/chemistry side of it is pretty useless. It doesn't make any difference whether you understand Newton's laws of gravity—gravity still exists whether you know about it or not.' 'Working day in, day out, to come to a conclusion which never seems to materialise.' The dangers of science were commented on by a small number. 'I may invent something which will eventually destroy the human race.' 'They' (scientists) 'can become the tool of a government, doing things not for an improvement but to turn the people to one particular party.' ' . . . it's a man's job, it's dangerous' (from a girl).

'I would like to marry a scientist because . . .' produced a few stereotypes of women scientists, but more strongly it evoked the egalitarian expectations which these teenagers have about marriage in general. Both sexes considered that they could learn from the scientist spouse, and could share their interests and appreciate their talents. The implicit assumption here is that both spouses would be scientists. Sometimes the appreciation was a little double-edged: 'She would be logical in her approach to life and would not be prone to irrational outbursts of temper.' 'He would be intelligent and quick-minded and could make decisions, I hope.' Both sexes appreciated the advantages of a spouse with a high income. Girls, however, were more likely than boys to regard their spouse's talents as having a use round the home, and girls, but not boys, thought that the fame and distinction of their spouse would be an asset. In other words, although these responses demonstrated an egalitarian view of the marriage partnership, the *material* advantages were expressed in implicitly conventional sex-role terms.

Objections to a scientist as spouse centred again on the negative image of the scientist as a person or type. The most frequent objection given, especially by girls, was the time-consuming nature of the job and the fact that it would take the spouse away. Both sexes regarded the narrowness and technical nature of the work as a disadvantage, and this was also expressed in terms of the kind of person who would go in for such work. For both sexes the supposed high intelligence of the scientist presented a possible threat to their self-esteem. The presumed lack of physical attractiveness, and non-normality, of the scientist of either sex was extensively commented upon. 'They seem too wrapped up in their work, only seem to bother with their experiments.' 'A scientist really has to be "married" to his job.' 'I'd get sick of his persistent braininess and his solitude.' 'She might be cleverer than me and show me up.' 'He would always bring up science when I am trying to talk about things in everyday life.' 'She would be too technical.'

This study confirms many of the components of the conventional image of the scientist. However, it is interesting that, in contrast with Mead and Métraux's findings, the scientist is now seen as quite *mundane*. The great achievements, helping mankind and so forth, are less the concern of these teenagers than the day-to-day pleasures and frustrations of the job.

The reality of the scientist

Is the scientist really the cold, irrational, unattractive man that students and schoolchildren alike describe? If the image of science is an accurate reflection of the role of the scientist and of scientific activity, then attempts to interest more girls in science must take a very different course than if the image is inaccurate. There have been a number of studies of scientists, mainly but not exclusively of eminent male scientists. Roe (1951) found that top scientists were happier with ideas and things than they were with personal relationships, but contrary to the stereotype, they were well balanced, stable family men. Studies of British and German engineers found that they were liberal and independent, but with a certain degree of puritanism. They saw themselves as practical, intelligent and self-assertive. These men are, therefore, not unlike the practical, unemotional, masculine person of the conventional image (Lawrence, 1977; Gerstl and Hutton, 1966).

In contrast, some studies of top creative scientists, those on the frontiers of their subject, showed a picture rather different from the stereotype. Mitroff studied *Apollo* scientists (Mitroff, 1974; Mitroff *et al.*, 1977). Mahoney (1976) and Eiduson (1962) also studied working research scientists. From these studies certain generalisations can be made. First, whatever the characteristics of the *stereotypical* scientist (the 'hypothetical scientist' which Mitroff's subjects easily identified) the requirements of the *actual* scientist are involvement, commitment to a particular point of view, the capacity for intuitive leaps or hunches, and the courage to take risks. This is in contrast with the supposed detached, objective, purely rational sifting of data which the stereotype suggests. Scientific progress requires rigorous *method* but also creative, imaginative speculation in development of theory and in problem solving. These characteristics are recognised and endorsed by practising scientists.

In terms of personality characteristics and behaviour there is also a gap between the real and the stereotypical scientist. Although it would seem that the scientist is rather cut off from family life, he does not work in isolation. Scientific activity is group activity, involving team work and considerable action with others. In this context, all the studies found that the scientists demonstrated considerable interpersonal aggression, jealousy and competitiveness, *but* that their emotional lives were bound up with the work situation, both the joys and the anguishes, rather than with their families. (A taste of this kind of interaction can be gained from Watson's *The Double Helix*, 1968.) It would appear from the reports that some scientists tried to run their emotional lives, *outside work*, more on the lines of the conventional image of the scientist.

Top scientists, then are highly talented people who are passionate in the work context, both about ideas and in their interactions. In this they do not fit the image, nor do they manifest the 'Roundhead' type of masculinity. In fact Mitroff argues that *stereotypical* masculine modes of thinking akin to the Roundhead style would inhibit progress. On the other hand, the degree of aggression demonstrated by these men is stereotypically masculine—even though the spite and jealousy have more conventionally 'feminine' connotations.

A major question is, how far are the characteristics attributed to the stereotypical or hypothetical scientist necessary for the pursuit of knowledge and scientific advance? In fact it would seem that the characteristics of aggression and competitiveness are not required of the scientist *per se*, but they are required of the successful person of either sex who wishes to compete in a world where such (masculine) qualities are valued in addition to basic ability. The most gifted in any profession are necessarily exceptional, and they, more than average exemplars of the profession, are likely to have adopted a life style which extends beyond the working day. The 'average' engineer or industrial chemist, may need for his job less intuition and capacity for imaginative leaps, but, paradoxically, more detachment and unbiased objectivity. On the other hand, his commitment to the scientific life style is likely to be less; it is a job like any other.

The difficulty with the stereotypical image of a scientist is that two images are confused. The boy or girl thinking of becoming a scientist is more likely to enter the life of the workaday scientist; yet their initial interests may have been kindled by the image of the top scientist. The gap between mad brilliance and dullness, and between the responses to 'When I think of a scientist I think of . . .'

and 'I would like to be a scientist because . . .' is some demonstration of this. If we want to encourage more girls to enter science it may be important to clarify this distinction, and to reveal as mythical the masculine connotations of the present stereotype.[5]

Notes

1 As I will show, the stereotype of the scientist is male; accordingly, I have used male pronouns for the scientist when discussing this stereotyped image.

2 The sixth-formers studied by Smithers and Collings (Chapter 12) thought the same: both male and female science specialists were considered to be more masculine than arts specialists of the same sex. [*Editor*]

3 Some of this material has already been published in Weinreich-Haste (1979) and Kelly and Weinreich-Haste (1979).

4 See the Statistical Appendix for a discussion of correlation and significance testing.

5 I am grateful to Clive Stevens for his assistance in coding the data in the second study, and to Stephen Cotgrove and Andrew Duff for directing me to some of the material on the sociology of science.

References

Beardslee, D. C., and D. D. O'Dowd (1961), 'The college student image of the scientist', *Science*, 133, 997.

Bendig, A. W., and P. T. Hountras (1958), 'College student stereotypes of the personality traits of research scientists', *J. Educational Psychology*, 49, 309.

Chetwynd, J., and O. Hartnett (eds.)(1978), *The Sex Role System*, London: Routledge & Kegan Paul.

Cotgrove, S. F. (1974), 'Technology, rationality and domination', *Science Studies*, 4.

Eiduson, B. T. (1962), *Scientists: their Psychological World*, New York: Basic Books.

Gerstl, J. E., and S. P. Hutton, (1966) *Engineers: the Anatomy of a Profession*, London: Tavistock.

Hoffman, L. W. (1972), 'Early childhood experiences and women's achievement motives', *J. Social Issues*, 28, 129–56.

Hudson, L. (1968), *Frames of Mind*, London: Methuen.

—(1970), 'The choice of Hercules', *Bulletin of the British Psychological Society*, 23, 287–92.

Kelly, A., and H. E. Weinreich-Haste (1979), 'Science is for girls?', *Women's Studies International Quarterly* Vol. 2, 275.

Lawrence, P. A. (1977), 'Engineers : the image and the reality', *Energy World*, June, 2–4.

Leiss, W. (1972), *The Domination of Nature*, New York : Braziller.

Maccoby, E. E., and C. N. Jacklin (1976), *The Psychology of Sex Differences*, London : Oxford University Press.

Mahoney, M. J. (1976), *The Scientist as Subject: the Psychological Imperative*, Ballinger.

Marcuse, H. (1964), *One-dimensional Man*, London: Routledge & Kegan Paul.

Mead, M., and R. Métraux (1957), 'Image of the scientist among high-school students', *Science*, 126, 384–9.

Merton, R. K. (1967), *Social Theory and Social Structure*, Free Press.

Mitroff, I. I. (1974), *The Subjective Side of Science: a Philosophical Inquiry into the Psychology of the Apollo Moon Scientists*, Amsterdam: Elsevier.

Mitroff, I. I., T. Jacob and E. T. Moore (1977), 'On the shoulders of the spouses of scientists', *Social Studies of Science*, 7, 303–27.

Roe, A. (1951), 'A psychological study of physical scientists', *Genetic Psychology Monographs*, 43, 121–239.

Rose, H. and S. (1969), *Science and Society*, Harmondsworth: Penguin Books.

—(eds.) (1976), *The Radicalisation of Science*, London: Macmillan.

Sellar, W. C.,and R. J. Yeatman (1930), *1066 and All That*, London: Methuen.

Tangri, S. S. (1972), 'Determinants of occupational role innovation among college women', *J. Social Issues*, 28, 177–99.

Tyler, L. E. (1964), 'The antecedents of two varieties of vocational interests', *Genetic Psychology Monographs*, 70, 177–227.

Watson, J. (1968), *The Double Helix*, London: Weidenfeld & Nicholson.

Weinreich-Haste, H. E. (1978a), 'Sex differences in "fear of success" among British students', *British J. Social and Clinical Psychology*, 17, 37–43.

—(1978b), 'Stereotyping, the sex factor', *Psychology Today*, June, 20–5.

—(1979), 'What sex is science?', in O. Hartnett, G. Boden and M. Fuller, (eds.) *Sex Role Stereotyping*, London: Tavistock.

PERSONAL EXPERIENCE

17

The pupils' viewpoint

This chapter is in two parts. First there is a collection of extracts from schoolgirls' accounts of their experience of science education. These accounts were obtained by advertisement. I placed notices in the newsletters of the Association for Science Education, Women and Education and the Women's Research and Resources Centre, asking women to write to me with their own recollections of school science, or, if teachers, to ask their pupils to write. A few individuals replied, but the greatest volume of responses came from three schools and a college of further education where the topic was set in lesson time. The comments should therefore not be taken as representative of schoolgirl opinion in the country. Although I have tried to include a variety of the views presented, other schoolgirls may well think quite differently. Indeed, it was noticeable that pupils from different schools wrote about different things. Some sets of essays described the conduct and content of their science lessons, while other sets concentrated on how the girls had made the decision whether or not to continue science. Clearly the way in which the teacher introduced the essay topic had been a strong influence on the subject matter of the essays. Whether it had influenced the views expressed is less clear. All that can be said with certainly is that all the sets of essays contained both positive and negative (often extremely negative) statements.

I eventually received approximately three hundred essays, ranging in length from two lines to two pages. From this wealth of material I have selected passages which are illustrative of some of the themes of this book. These are not necessarily the most logical or most eloquent passages. But they do convey some of the flavour of the essays, and they indicate how some of the abstract ideas contained in previous chapters are manifested in the classroom. It is in this light that they should be read: as illuminating and complementing the more rigorous studies.

The second part of the chapter, 'Girls, physics and sexism', is an edited transcript of a conversation between two schoolgirls. 'Mani' and 'Donna' are in the fifth form of a mixed comprehensive school in London. 'Mani' was born in Bangladesh and came to England when she was six. 'Donna' was born in England, but her mother is from Sierra Leone. Both are currently attempting several subjects, including physics, at CSE. During their fourth year the head of science at the school, a feminist, realised that these two girls were having difficulties in physics. Although she did not actually teach them, she took particular interest in the progress of girls studying science, and so she arranged to meet these girls. It soon became apparent that there were two separate but related aspects to their problems. They were having real difficulty with some of the content of the course—the terminology, the maths, the unfamiliar and abstract problems. But they were also feeling very uncomfortable on account of the sexism that dominated classroom interactions in physics. They were the only girls in a class with a male teacher. Both the boys and the teacher were overtly sexist in ways which embarrassed the girls and undermined their confidence, with the result that they were unable to admit that they did not understand some of the work.

The two girls continued to meet the woman science teacher fairly regularly once a week for nearly a year. At their meetings they talked about the position of women and feminism as well as doing some physics. During these discussions the girls became increasingly aware of sexism as an identifiable form of prejudice, and increasingly determined to fight it. When made aware of the opportunity to contribute to this book, they were keen to record their experience. The transcript in the second part of this chapter is the result. [*Editor*].

From schoolgirls' essays

Isolated and ignored

I was in a class of 32 pupils all of which were boys except me. I used to get avoided and the lesson would revolve around the boys in the class, and I would be one of the lads. I was counted out when experiments were done, but put upon when home work should have been handed in. It would be then that I was noticed—*Sixteen-year-old school leaver, mixed comprehensive school.*

Two girls were sitting in the middle of the class and there'd be two rows of boys either side and the teacher would always direct his questions to either side of the classroom avoiding us in the middle and so Lizzie and I used to score points every time we'd answer questions we used to mark it on the desk, you know one for us, one for them.—*Lower sixth-former, mixed comprehensive school.*[1]

If we did an experiment it would always be the boys who would do them, the teacher would say that the boys should do the experiment because he must have thought the girls were stupid, he would say 1 girl to 3 boys and do the experiment but it would be the boys who did it and the girls who watched.—*Sixteen-year-old school leaver, mixed comprehensive school.*

I didn't think anything about this bias before I took chemistry but now I'm the only girl in the class I find that at first I'm not really expected to understand so even before we've approached the subject I think well I'm not going to be able to understand it. My sister had done maths and she said that in the maths class there were two girls who sat at the back and all the boys got everything first time and when she went for extra help the teacher said 'oh well I don't expect you to understand it so just get the formula learnt' and so this is what I'd expected. In fact I got a higher grade at O level than any of the boys so, we don't get any talk about women's inferiority now.—*Lower sixth-former, mixed comprehensive school.*[1]

The reason why I don't like science is because there was only a few girls in the group and the rest was all boys. And the science teacher was always telling us off in front of the lads and she never seemed to tell the lads off. Or if any of the girls was going out with one of the lads and was sat with each other she would come and split them up for no reason at all.—*Sixteen-year-old school leaver, mixed comprehensive school.*

The Headmaster had several discussions with me, aimed at dissuading me from taking four A levels, all in science (although several arts students were taking four arts subjects to A levels without any queries being raised). I remember him saying words to the effect that it was unbalanced for a girl to do all science subjects and that if I became a career girl and failed to have children then I was failing in my duties as a woman. My O level Physics teacher had encouraged me to continue with Physics to A level but was not herself qualified to teach to that level and our school did not have the facilities. Consequently, I attended a nearby boys public school for my Physics lessons. This was extremely unsettling for me in several ways. Firstly, I had to be ferried between the two schools thus often arriving late for lessons. Secondly, I came from a totally female environment to a school of 600 boys and I was the only girl. I became very self-conscious. Thirdly, one of the Physics teachers was obviously embarrassed at my presence in his classroom and I was rarely encouraged to speak, question or even attempt practical work – I was always 'helped' if anything vaguely technical was involved. In my Physics classes I became very inhibited and withdrawn. I did not make my difficulties with the subject known and I failed my A level, only getting an O pass. I harbour considerable resentment towards my school for its lack of science facilities and inadequate science education in the sixth form.—*Biological science graduate, now twenty-three, girls' independent school.*

I found the girls to be more studious than the boys, but as time went on the boys were encouraged more and the girls were not actually put down but were, shall we say dampened.—*Sixteen-year-old school leaver, mixed comprehensive school.*

A boys' subject

I have always associated chemistry and physics with boys and I've always associated biology with girls. So when I thought of chemistry I thought well boys, mostly, so I just scored it out. I don't see why a girl shouldn't be able to do chemistry and physics. I'd liked to have done them although I really wouldn't have been very good at them. I'd like to have carried on with them but as I said before, boys' subjects, so I just dropped them.—*Lower sixth-former, mixed comprehensive school.*

I think girls find it more difficult to be interested in Physics and Chemistry, after all you don't hear of many lady teachers in these subjects. But Biology is more interesting and there are quite a few lady teachers. It is still more for boys than girls even if there is the Sex Discrimination Act.—*Fourth-former, girls comprehensive school.*

We were great friends with the boys. Most of us weren't as brainy as the boys though.—*Sixteen-year-old school leaver, mixed comprehensive school.*

Biology is the best science for girls because you understand it better.—*Fourth-former, mixed comprehensive school.*

Fighting back

With a lot of argument and bad feeling, my parents and I discussed if I should drop languages and take up sciences. My father, a typical chauvanist said that girls should study languages and not sciences. I disagreed very strongly, because of my liberated nature, and finally I won through. I consider it very important for girls of my disposition, growing up to a technical future to understand science. I consider physics to be a majorly important subject due to the fact that my future is to be surrounded with technology. I am interested, quite genuinely, in mechanics and in my adult life I see no reason why I being female, shouldn't be able to fit a plug, mend a radio, etc., just as well as any man if not better. I am infuriated when men cast off all hopes of women being scientific because women are liberated at last.—*Fourth-former, girls' comprehensive school.*

I think that sciences are important for everyone to have, especially girls, because more women need to go into the scientific side of work, mostly to help the world and partly to help their kind (e.g. women.). Sciences are considered generally unimportant in girls' schools by the Education Authority, some teachers and even parents and pupils. As I believe that girls need sciences (not including domestic science) as much as boys, that swung my decision to take biology and physics a great deal.—*Fourth-former, girls comprehensive school.*

I was in a slightly different situation because I went to an all girls' school – a boarding school – I only came here for my sixth form, for my 'A' levels and consequently there was no judgement, no thought of whether there'd be any more boys or more girls in the classes. I chose it because it was the subject that I could do. I must admit I was a bit surprised when it turned up that there *were* only three girls in my chemistry and about thirteen in the group altogether and the boys expect you to be the three lowest in the class. I'd never been in a mixed school since I was at junior school, and I'd sort of never really realized that you were expected to do cookery and the arts rather than the sciences.—*Lower sixth-former, mixed comprehensive school.*[1]

Puberty

Another reason why I didn't really try is because there was a boy called Tom in our science class. I was at the age when every school girl has a heart throb. During science I would sit at his table and just watch his every move. I used to get told off for not listening to the teacher but I didn't mind because it showed Tom I was not a 'swot'. I was more interested in what Tom was saying to me, rather than what the teacher was saying. I think it would have helped if there had been no boys in the science class that we were in. The boys always seemed better at science and understood the equations in chemistry. To me equations were uninteresting and I couldn't at the time see their use.—*Sixteen-year-old school leaver, mixed comprehensive school.*

When I went to my senior school and had our first few science lessons, I thought it was great, but that soon ended. The teacher, being a male, started to show interest in one girl, gave her extra home work, and often saw her at playtimes. Nearing the end of the first year, the teacher started to get more friendly with a few other girls, never boys, and of course because he hardly ever paid attention to the boys they lost interest, and started playing about, and when they did that they got the slipper for very petty reasons. He persuaded my parents that I could take Chemistry at 'O' level, and like a fool I did. It was then I realised what he was after, he started to look down the girls' blouses and tickle them, and drop things down their blouses. I never told my parents because they would have gone mad.—*Seventeen-year-old school leaver, mixed comprehensive school.*

In the fourth year I took up a subject called Human Biology. This got down to even more detail about the human body but the girls found it hard to communicate to the boys because they were embarrassed, the boys always thought they knew it all, so our class was often found to be quiet.—*Sixteen-year-old school leaver, mixed comprehensive school.*

Likes and dislikes

Our teacher would always make human biology interesting but plant biology seemed pointless. I mean to say who needs to know about how a plant breeds or germinates. But learning how the human body works was certainly interesting. Biology was the only science I took. I hated doing experiments with plants too, everyone would more or less get the same results. But the human experiments like skin under the microscope, etc., would all be slightly different.—*Sixteen-year-old school leaver, mixed comprehensive school.*

Chemistry I thought was a wasted subject. The reason I think it was wasted is because the gases and formulas are useless to a person unless they want to study the subject in great detail. When being taught the subject I found it boring and would rather study something much more worthwhile to be put to use for my final career. The only science that I carried on through my full education at secondary school was Domestic Science which I did enjoy. The science that I would've enjoyed would have been biology (human) but again, no human biology, when the subject was compulsory, was taught. They taught me all about plants and prehistoric animals which I disliked immensely, so this deterred me from taking this subject as an exam.—*Sixteen-year-old school leaver, mixed comprehensive school.*

I chose Biology because it seemed to cover more areas of life whereas in Chemistry and Physics it seemed to be only pouring liquids, powders and gases into test tubes which didn't really appeal to me. Biology I thought was most important because I found out about things which could be of help to me in the future, e.g. correct foods to eat, how it all works in the body, about breathing, how plants and animals live, etc., which was very interesting for me. But I could not understand much reason to know whether one liquid was heavier than the other, etc.—*Fourth-former, girls' comprehensive school.*

The work in science never appeared to relate to the environment in which I lived and science seemed to me unworthwhile as the answers were already known and easily available without the practical work.—*Trainee teacher (now twenty) girls' grammar school.*

I enjoyed these lessons, especially chemistry. It was good to be able to experiment and mix chemicals together.—*Sixteen-year-old school leaver, mixed comprehensive school.*

Doing all those experiments were just a bore. If the great scientists of today says something, I will take their word for it until someone else proves that it is wrong.—*Fourth-former, girls' comprehensive school.*

For the first two years, I did enjoy general science, the lessons were varied and the experiments fun – and time wasting. I always remember doing varied tests on the amount of chlorophyll in plants – using blotting paper, and ink dyes – the work seemed exciting – and I could cope, which helped a great deal.—*Eighteen-year-old school leaver (now thirty) girls' grammar school.*

Half the time, the experiments we did in chemistry, nearly everyone did not know what they were supposed to be doing. Most of the time we were just burning things in test tubes.—*Fourth-former, girls' comprehensive school.*

I dropped science for the simple reason of not understanding it. I managed quite well with it in the third year but I never understood it. It was all very well being able to draw neat diagrams and coloured pictures of little stick men holding buckets of water and metal rods, but I knew that when it came to taking exams, I would not get very far on pretty pictures. When I see the books now that people who took the sciences are working from I realise I was right.—*Fourth-former, girls' comprehensive school.*

I took physics as I had more chance of passing it than chemistry or biology, and I was interested in motor bikes (physics helped me understand how the engine worked.) I found physics quite easy as it involved a lot of maths, and I quite liked maths and was OK at it.—*Sixteen-year-old school leaver, mixed comprehensive school.*

Chemistry was a little easier than physics and I decided to attempt 'O' level. It was decidedly more interesting and I enjoyed the experiments, and having an excellent teacher (male) for this did help. But the crunch came when we had to do calculations – a large proportion of maths, and this was when I knew I would fail – miserably. No matter how many extra periods, the teacher would spend with me, I just panicked, and resolved to the fact of failing.—*Eighteen-year-old school leaver (now thirty), girls' grammar school.*

For the first two years we did a boring subject about electricity which didn't really appeal to any of the girls in our class. The teacher took no interest in the girls what so ever. He cared about the boys' future and not the girls.—*Sixteen-year-old school leaver, mixed comprehensive school.*

I took Physics because I really enjoy the subject, and I like playing around with lenses and raylamps and all the experiments that we do.—*Fourth-former, girls' comprehensive school.*

Squeamish?

I gave up Chemistry because I hate the smell. Every time I went in the Chemistry lab I felt sick.—*Fourth-former, girls' comprehensive school.*

I hated biology. I do not think I could stand seeing all those small animals cut up and used in experiments so I'd rather forget that subject altogether.—*Fourth-former, girls' comprehensive school.*

Sometimes we would get the boys to come and help us if it was something like dissecting a rat or spitting into a test tube.—*Sixteen-year-old school leaver, mixed comprehensive school.*

He always seemed to pick on me or my friend to remove the dismembered carcase of a poor innocent mouse or locust. This definitely put me off biology.—*Sixteen-year-old school leaver, mixed comprehensive school.*

Teachers

The first year of the course was very interesting. What in fact made the lessons was the teacher, he had a fantastic way of conveying everything you were learning, nothing seemed very hard to learn.—*Sixteen-year-old school leaver, mixed comprehensive school.*

My third year was a nightmare. My one failing was Physics, which I could never master – and having a hopeless teacher didn't help. He distinctly hated females, and made every effort to make this fact known. I fumbled through that year, and made a resounding resolution never to do Physics again.—*Eighteen-year-old school leaver (now thirty), girls' grammar school.*

He would fool around and talk about football with the lads in our class (as often happens with most subjects).—*Sixteen-year-old school leaver, mixed comprehensive school.*

As a child I enjoyed science and did well in my science classes. Any ambitions I ever had for a career as a scientist, however, were very firmly dashed when I was 14. While in the lab after school the science teacher and I discussed a problem I had encountered in a book he had loaned me. After we had worked on this for a while he commented that he was very impressed with my ability since the book was one which he had used as a text in university. He also commended me on producing the very best end-of-term exam paper and the highest marks in the class for my lab work, and advised me to follow up on my scientific interests and my superior achievements by going on to a career in teaching. He then went on to talk about a boy in my class, commenting that while he was not nearly as good a student as I was, he really ought to be encouraged to pursue a career as a research scientist. I did not question his judgement and left the class feeling that this boy's scientific brilliance (which was completely invisible to me) was a reality which could be detected by this experienced teacher. In place of the intense curiosity and excitement of discovery which had motivated me in most of my classes, I was left with a vision of myself as a dull plodding grind, not a scientist. That vision endured for many, many years. I went on to a career in art, married a scientist, then took a degree in social science, and, in my middle thirties, a Ph.D. in psychology. That teacher cost me over 20 years as a research scientist.—*University lecturer in psychology (now forty), mixed grammar school.*

Timetabling

Whilst filling in my option sheet I came across a problem. I either had to drop German and carry on with all three sciences or carry on with German and take one. I carried on with German and chose Physics. Physics appealed to me because it comes in useful for minor things such as changing a plug, and when dad is not around I take over as chief electrician.—*Fourth-former, girls comprehensive school.*

I did not want to take up Physical Secondary Science, I wanted to do Chemistry, but we only had the choice of Biology, or Physical Secondary Science. Whereas band one had the choice of Biology, Astronomy, Physics, Chemistry, Geology and a few others. As we were not allowed to do the other subjects I go down town to the Central Library, and teach myself Geology by reading books on it and doing exercises out of some books. I find it very interesting.—*Fourth-former, girls comprehensive school.*

I was going to take Physics with Chemistry at first but as not enough people wanted this option I did Biology instead.—*Fourth-former, girls comprehensive school.*

I gave up sciences because I preferred languages and wasn't able to take any science at all if I did three languages, all of which I preferred to any science.—*Lower sixth-former, girls' independent school.*

I chose Biology as a science class because I wasn't good enough to take a language.—*Fourth-former, mixed comprehensive school.*

I didn't mind doing biology because it isn't so much of a science. I took chemistry because there wasn't anything else for me to take so I did it to fill in my timetable.—*Fourth-former, girls' comprehensive school.*

I opted out of science mainly because in the columns was other subjects that I wanted to take.—*Fourth-former, mixed comprehensive school.*

We had to do at least one science at exam level and so I chose to do biology, because someone told me it was the easiest out of the three.—*Sixteen-year-old school leaver, mixed comprehensive school.*

Exams

When it came to the end of the third year and option choices I wasn't too sure whether or not to take chemistry as I didn't know whether I'd be good enough for 'O' level so the teacher told me he could put me in a mixed C.S.E. and 'O' level group and make my mind up nearer the end of the course. This is what I did! A few months before exam time I lost my nerve a little and plumped for C.S.E., but I got a C.S.E. Grade 1, so I didn't do too bad.—*Seventeen-year-old school leaver, girls' comprehensive school.*

There was a choice of whether you took Nuffield biology which was mainly about plants and animals or human biology. It was very unfortunate though because the human biology course was a CSE course and the Nuffield was an 'O' level course.—*Seventeen-year-old school leaver, girls' comprehensive school.*

In the 4th year I did Human Biology and had a female teacher. She was really nice but gave us loads of work. It was very interesting and has always remained my favourite subject. At the end of the 5th year she wouldn't allow us take 'O' level and so we all had to take C.S.E.—*Sixteen-year-old school leaver, mixed comprehensive school.*

I was recommended for 'O' level Physics along with many other girls in the class, but our teacher put us off, 'You don't want to take Physics, it's a boys' subject,' he said. 'You'll find it very boring and difficult.' So even though the majority of the girls enjoyed Physics only one girl in the class finally took the subject through to 'O' level standard.—*Sixteen-year-old school leaver, mixed comprehensive school.*

Science or arts?

I have always favoured the science-based subjects, finding them more logical and containing a larger degree of common-sense than, for example, English. The sciences are not a question of learning facts, but more of understanding the underlying principles.—*Sixth-former, girls' grammar school.*

Both Physics and Chemistry were taught *not* on a conceptual basis, as if we could *not* understand and enjoy the ideas and principles of science. We were taught science as if it was learnt (like other subjects) by taking down endless facts and notes. Yet for me, the initial attraction of science (and maths) was that the learning process was less tedious.—*Eighteen-year-old school leaver (now thirty-five) girls' Church school.*

My mother, who left school at 14 and left work on marriage, was a purely domestic person. I, as a 'bright' child doing well at school, was obviously going to have a life different from hers, but my only role model was my father, a physics graduate and scientific civil servant. It was assumed without argument that I would be like him. In retrospect I can see that I was an exceptionally docile, maleable child with a strong respect for authority and no strong sense of my own identity. I had no real image of what my future life would be like and no positive engagement with any of my school work. So being like father was a weak reason which prevailed in the absence of others. When I had to choose my O level subjects I naturally, therefore, made a science-based selection. At this point, to my great surprise, I received an emotional appeal from my English teacher not to 'give up' my 'best' subject i.e. English, which I happened to get highest marks for in exams (I was in fact not 'giving it up' but by specializing in science precluded the possibility of making it my speciality later). Stunned by the evidence that anyone should care so much, I went home and told my parents that Miss X thought I should do English as I was 'better' at that. I don't remember having any strong feelings either way. My father has since told me that he was very disappointed, but believing that I should have the right to choose, left the decision to me. I therefore followed what had been the strongest (not most logical) expression of opinion, dropped all science subjects except maths, and opted for a straight arts programme. I can't say that I have since regretted the decision, but I am often appalled at the haphazard way in which it came about. I think the determining factor was the lack of role models. Apart from my father, the *only* people I knew with professional careers were my teachers. Teaching science and teaching arts seemed to me much of a muchness, I suppose. If I had been acquainted with a greater variety of possibilities, I might have made a more informed choice.—*University lecturer in English (now thirty-seven), girls' grammar school.*

When I studied sciences I enjoyed the subjects very much. I did not carry them on because I didn't have the necessary brain power.—*Lower sixth-former, girls' independent school.*

When I decided at age 14 that I wanted to be a chemist, and was being 'fought over' by various subject specialists in the school, my grey-haired headmistress said over the top of her pince-nez 'Are you more interested in people or things?' I didn't see the point at the time, but I think, basically, that remark is very important.—*Head of science (now forty-two), teacher training college.*

Careers

I chose to keep up all three sciences because of the type of career I want after I leave school. I think that there is a wider choice of interesting careers if you have the ability to work computers and know what they are made of. There are many types of careers in hospital, for which you need sciences, which especially appeal to me.—*Fourth-former, girls' comprehensive school.*

I chose to keep going with the three sciences, Biology, Physics and Chemistry. My main reason was that I wanted to do something like Architecture or Engineering because I feel that there is a whole new field of jobs opening up in these areas for young women and men to do. As I enjoy science and design I thought that there would be interesting jobs available if I worked and gained the correct qualifications.—*Fourth-former, girls' comprehensive school.*

I took chemistry because I need at least two sciences. I do not like the subject very much but kept it up so it would help with my nursing career as I would have a better chance of getting into a hospital with two sciences.—*Fourth-former, girls' comprehensive school.*

I took geography, biology and physics because I like these subjects and they interest me, also I hope to become a radiographer and these subjects are essential if I am to pursue this career. I did not take chemistry because, it did not seem essential in my choice of career although now I have realised it would have been better to take it.—*Fourth-former, girls' comprehensive school.*

I wish to be a hairdresser and I was told that you need to have physics, biology and chemistry. To be a hairdresser I would need biology to know about the skin and your hair, chemistry to know about mixing chemicals, and physics to know how to mend a plug on a hair dryer or even mend a hair dryer. At times in these subjects, however I wish I hadn't chosen these subjects because I don't understand a thing the teacher is gabbling on about.—*Fourth-former, girls' comprehensive school.*

The reason I took science was because I thought it would be a good basis for my hairdressing, there was no other reason because I detest the lesson.—*Sixteen-year-old school leaver, mixed comprehensive school.*

In 1932, I was 14 years old and in my first year at a girls' private boarding school. My father and the school explained, it would be best for me to take Domestic Science rather than Chemistry and Physics. I was the only girl of the family and as my mother had recently been diagnosed as having multiple sclerosis, it was thought I might have to keep house for my father; also I might marry and not have a domestic servant. However, in my twenties, I successfully studied Anatomy, Physiology, Midwifery, etc., and had a nursing career of more than 10 years.—*Seventeen-year-old school leaver (now retired), girls' independent school.*

I kept up Physics because I thought it was an interesting subject and might help me with electronics when I own my own house and a fuse needs changing, or a new plug needs to be put on an iron.—*Fourth-former, girls' comprehensive school.*

Science will not help me being a receptionist and is a total boring waste of time. I could already fix a plug without being told.—*Fourth-former, girls' comprehensive school.*

Parents and others

My choice in taking all three sciences was maybe because of my brother. He has taught me to enjoy science subjects very much or maybe because my parents wanted me to go into a medical career.—*Fourth-former, girls' comprehensive school.*

A lot of my friends took science and I wouldn't like to be separated from them all the time.—*Fourth-former, mixed comprehensive school.*

I chose physics not because I like it, mostly because my parents wanted me to.—*Fourth-former, girls' comprehensive school.*

In chemistry I like the experiments but am sorry we do the modern course as my father did traditional chemistry and so can not help me much. Physics is my worst science and I also wish we did traditional physics as the modern method of 'find it out for yourself' is muddling.—*Fourth-former, girls' comprehensive school.*

Girls, physics and sexism

Mani: Our fourth-year Physics class was dominated by boys and I found it very difficult to feel a part of it.

Donna: Yes, particularly with the boys' conversation. They always talk about, you know, dirty things about girls and men, and usually they talk about girls like they're just a sex symbol, which is really embarrassing when there's two girls and you feel you're being discussed. That's what really made me hate Physics, was that they talked about some embarrassing things and they made you sick.

Mani: And the teacher doesn't say anything to them, you know he'll let them carry on.

Donna: It's very strange, really, because that teacher is very strongly opposed to racism.

Mani: Yes. But he didn't seem to mind about the sexist remarks. Teachers shouldn't allow conversation like that. You know, if a boy is carrying on a conversation like that I think he should be stopped.

Donna: Yes, like he does with racist jokes. It's the same thing, really. Because it's discriminating against girls and so on. This racist business—I mean, you feel inferior to another race.

Mani: But if you talk to them, tell them what you feel about men doing things like that, then you can get a few things out of them. You see, a lot of the boys don't really agree with these remarks.

Donna: But there were other ways that I felt left out as well, like the boys would say that girls aren't good at anything.

Mani: Yes, they'd say that men are better than girls, and girls shouldn't be doing Physics. Physics is really a boys' subject. And they'd say this straight to our faces.

Donna: That really used to make me angry, and so I started arguing with them.

Mani: The teacher didn't really help the situation, because when the boys were messing about he used to say things like 'Stop laughing like little girls'. Teachers really shouldn't say that sort of thing.

Donna: Yes, that makes me feel really low. And especially when there's only two girls.

Mani: We were treated differently; he would say things and make comments about girls to boys. And we couldn't say anything because there were only two girls, so we couldn't just shout out. Or if we were having a conversation, he would say something like 'This is not a Women's Institute'.

Donna: Yes, if we were talking he'd tell us to be quiet, but if the boys were talking he didn't mind. Suddenly he'd come up and tell them to be quiet, but not as much as he told us. Maybe we felt that way because we are aware of it and maybe it isn't really that he's picking on us: we might just think that he is because we've got it in our mind.

Mani: But none of this happened in the other two science subjects. Maybe because we had a female teacher for both Biology and Chemistry.

Donna: And there's a better mixture of girls and boys.

Mani: Yes, it was really nice. You're not scared that if you get an answer or something wrong anybody's going to go and laugh at you, because the teacher won't allow that. If we do it in Physics they burst out laughing.

Donna: The same boys who are in Physics, they talk quite differently when they're in Biology or Chemistry.

Mani: Yes, two boys in particular behave differently in Physics. In Chemistry and Biology they never talk about sex or anything. But maybe because there are more boys in Physics they find it easier; if there were a lot of girls I don't think they'd talk that much about these sort of things. There's another boy, he doesn't really seem to like male chauvinism, but still he makes these sexist remarks in Physics. I don't know if he means this for a joke or whether he really feels that way. But even jokes can make you feel really uncomfortable.

Donna: I think they behave like this because they were all brought up to believe that men are better than women. It's a role that society has more or less put on us, that girls do one sort of thing and boys do another sort of thing. You know, you only associate this type of work with a boy and associate this type of work with a girl.

Mani: I find that in our custom, the Asian custom, girls are treated very low and the boys are always put higher than the woman. The women are brought up to believe that boys are better, the mothers, the parents, teach them this. In our country girls don't go to work, whereas girls should go to work, and girls don't go to the market, either. They never go shopping for clothes or food. The men are supposed to go shopping, and the women stay at home and just cook the dinner for the husband, and that's all.

Donna: Yes, but now it's changing. In the village you will find all girls are treated like that but if you go to the town both sexes are nearly treated equally.

Mani: Not exactly equal. In some ways equal but in other ways it is not, because there's still the boys and the men sort of somehow or other coming to believe that they're better than the women.

Donna: That's strange, really, because your father is really ambitious for you, isn't he? I mean, he really wants you to do well at school.

Mani: Yes, but I think sort of he wants me to grow up and be educated, but then again he wants me to be like a lady, stay at home and be like a lady. He wants both, really, but I can't do both.

Donna: In my country we find that women can't get as many jobs as a man can, they can't get into higher-paid work, like a man. You won't find girls doing higher education, they usually stop going to school when they're in class 9 when they're about thirteen or fourteen or something like that.

Mani: But equal rights has helped a lot, where boys do Home Economics and girls can do Woodwork, things like that are the first step. We didn't have that in the first year, did we? When we came, we couldn't do Technical Drawing because it was for boys.

Donna: But now it is changed since the new first-year came. The boys can do Home Economics, girls can do Technical Drawing.

Mani: It often seems to work only one way round, though. Only about one girl does Woodwork but quite a lot of the boys do Home Economics.

Donna: Even so I wouldn't like to be in a girls' school, I prefer a mixed school.

Mani: Yes, I want to be equal with the boys. I mean, why should I be away from the boys? I'm not afraid of them. I like to be with them. I think separate schools for boys and girls won't help the situation.

Donna: Boys going to boys' schools and girls going to girls' schools: I think it makes it worse, because then the girls begin to hate boys more and the boys begin to hate girls, because they think they're in a society of their own. In fact in this generation I don't really believe in separate schools for boys and girls. I prefer mixed schools.

Mani: Yes, even though we have all these problems with the boys this is the only way, by doing Physics with the boys, that's the only way we're going to overcome the situation.

Donna: I used to think that mixed schools were horrible because the boys were silly and naughty and all that, that's why I didn't want to come to a mixed school. But now I think they're much better than girls' schools because you can mix with them, you can understand them, talk to them. They're OK, really, when you talk to them. And after all, when you have a job you're not going to have a job with all women, there's going to be men around.

Mani: I think it's beginning to change now; now we're in the fifth form it's much better. The boys act more their age, they're more sensible.

Donna: In fourth year they only used to talk about their own things, where they'd been, the pictures, X films—you know. But now they don't talk much about that. They talk about their work and they talk to us. They don't really make so much distinction between us, us being girls and them being boys.

Mani: Yes, it's much better, even in Physics. They're not all that noisy, they get on with their work, don't they? We really have helped the situation, though, by just facing up to it, come what may! Maybe because we answered them back and talked to them, they see they can answer us back and talk to us.

Donna: Yes, if we had been very quiet they might not have changed so much in their attitude towards us. It was important that we tried to actually shout out and talk to them. And try to mix.

Note

1. I am grateful to Mary Prichard and the BBC for permission to reproduce these comments.

18

The teachers' viewpoint

In this chapter we look at girls' performance in science from the teachers' point of view. In the first part Judy Samuel, a feminist chemistry teacher, writes in some detail about her experience of teaching in a mixed comprehensive school. Then follows a collection of extracts from teachers' letters, sent in reply to an advertisement in the newsletter of the Association for Science Education which asked teachers to write to me giving their opinions on girls' under-achievement in science and what if anything could be done about it. The extracts reprinted here are not a representative selection of the replies received, but are chosen to illustrate particular points of view or because they make suggestions which other teachers may find useful. I make no apology for this – the replies presumably came from teachers who were particularly interested in the problems of girls in science, and there is no special merit in presenting a representative selection of opinions from an unrepresentative sample. A surprisingly high proportion of the replies were from teachers in girls' independent schools. The view that boys are innately better suited than girls to science received very little support from these teachers, but differences in intellectual style between the sexes were frequently cited. The most popular explanations were in terms of poorer facilities in girls' schools, few role models for girls in science and little encouragement from home. Several of the writers developed quite sophisticated models in terms of the masculinity of science and girls' greater need for reassurance and encouragement, some of which are presented here. [*Editor*]

Feminism and science teaching: some classroom observations

by Judy Samuel

I have been teaching chemistry at Turton High School in Bolton for four years,[1] I describe myself as a feminist, and have been actively involved in the Women's Liberation Movement since before I began teaching. As a direct result of this personal commitment I am concerned about the incontrovertible fact that, particularly in mixed schools, fewer girls than boys study physics and chemistry (see Table 1.1; see also DES, 1975). The national trends are replicated in my school, and in this chapter I will recount some of the situations and incidents I have observed in my classes which may provide clues as to how this pattern arises. Whenever possible I have attempted to teach in a way that does not discriminate against girls, and I will try to indicate what a sympathetic teacher can do to encourage girls in physical science. But there are many factors beyond the control of an individual teacher, which operate within the classroom to influence the attitudes of students to physics and chemistry, and I will also try to show some of the constraints that these factors impose.

Turton High School is a mixed comprehensive, taking its 1,800 pupils from a large catchment area including both urban and rural districts. I teach chemistry throughout the school, but as I see a great many parallels between chemistry and physics I will try in this chapter to discuss both subjects and to point out the similarities and differences between them. I will not, however, attempt to deal with the situation in biology, which, it seems to me, is quite different. Nor will I discuss the sixth form in any detail: the proportions of girls in the A level science groups reflect fairly accurately the proportions in the O level groups from which they came. I have the impression that girls at Turton are not dropping out of science at this stage and that they eventually do as well at A level as their male counterparts.

All pupils at the school study physics, chemistry and biology separately in the first three years. In the first year they are taught in tutor groups (mixed ability), and are divided into A, B and C sets at the end of the first year on the basis of their performance.[2] The setting is reviewed at the end of the second year. In each science, all groups follow a common curriculum to the end of the third year: it is the depth of treatment which varies from set to set. Worksheets (devised by the science department) are used in all three sciences in the first year; they are also used in chemistry in the second year and in physics in the second and third years. I will deal with the options system used with the fourth and fifth years later.

When the new first-years arrive in August, many of them are rather 'awed' by the experience of being in a laboratory: if they have done any formal or systematic science before it will usually have been in their own classrooms. This awe rapidly disappears as we get down to work, but something else emerges: the different ways in which the majority of girls and the majority of boys react to the demands made of them. A simple example will illustrate what I mean.

A very early lesson is concerned with how to use the Bunsen burner. As soon as mention is made of what we are going to do, many boys comment that they've 'got one in their chemistry set', or they've 'seen one before', yet few girls make any comment at all. When the equipment is out, the girls often show considerable fear; for example, they shy away when their Bunsen is lit, and are unwilling to turn the collar to alter the type of flame. They have to be reassured that the collar is not hot before they will touch it. Whatever boys may feel at this stage, they very rarely show any hesitation in carrying out this experiment, and often have to be disuaded from 'overdoing it', whereas many girls stop as soon as they have done the minimum required.

Why do the girls so often react in this way? Many of them are undoubtedly familiar with using a gas cooker, and presumably the majority are used to handling hot pans and dishes when cooking, or see this as a usual thing for a woman to do. Is it the laboratory situation which operates against the girls, or their perception of science, or their unfamiliarity with the equipment? Or all three? Have the girls already been socialised into expecting teachers to expect them to be hesitant?[3]

Similar situations arise on other occasions during the year, when, for example, hot apparatus has to be handled or when acids are encountered for the first time. Few boys appear worried by having to handle hot apparatus, and often the opposite is true: a boy will pick up the apparatus even if he burns his fingers in the process. Often, too, I have observed boys within a group encouraging each other, or one of their number, to pick up the hot object. In contrast, a group of girls can often be observed all refusing to attempt to lift the object. They need to be shown the best or safest way to do it, and then watched and encouraged while they lift it themselves.

It seems likely that the differing experiences of girls and boys, at home and in the primary school, may contribute to these different patterns of behaviour in science lessons. Boys are often encouraged to play with mechanical, electrical or constructional toys and to help with tasks around the home involving tools, but girls are less likely to have this background experience.[4] Thus the girls are doing something new and unfamiliar in science laboratory classes, and it is, perhaps, not surprising that they often look for reassurance and encouragement, even though they have been given, and have understood, the directions. Yet if I, the teacher, did not watch for this and boost the girls' self-confidence where necessary, they would on occasion sit and look at their equipment instead of getting on with the experiment.

This discussion may appear to show first-year girls in physics and chemistry lessons as rather negative in their approach compared with the boys. Let me put the record straight: the first-year boys are far from perfect – they are sometimes carried away with enthusiasm, which may lead to accidents or breakages, and their energies often need to be channelled in more acceptable directions; for example, when using acids and indicators boys are often keen to 'mix everything together' in the hope that something spectacular will result. And the girls are quite positive in their approach to the subject overall: they are interested and keen to take up the challenge of a totally new area of study. They certainly do not, collectively or individually, dismiss it as 'not for them' at this stage, even though many of them do not receive much support or encouragement from their parents. (At parents' evenings I have sometimes had both mothers and fathers, usually of girls, make comments to me along the lines of 'Nothing personal to you, Miss, but don't you think that chemistry is more a subject for boys than girls?') Despite all this, and whatever disadvantages there may have been in their education up to the age of eleven, girls do achieve as well overall as boys at physics and chemistry in the first year.

By the time the students are in the middle of their third year the situation has changed considerably. Attitudes to science have become more polarised, and considerable differences have emerged between the attitudes of the high-achieving girls and the other girls.[5]

In an A set, the majority of the boys are openly enthusiastic and are working well, orally at any rate. Although girls and boys are equally represented in A sets, only some of the girls resemble the boys in their approach. These are the girls who enjoy science and see their future development and careers in this field; some have identified particular career aspirations such as medicine or engineering. Clearly these girls are as highly motivated as their male peers, and in my experience they do go on to do well at O level and A level and to continue to study for their chosen career at university.

But others among this group of able girls have lost much of their earlier interest. Although their written work is usually still of an acceptable standard there is little enthusiasm or interest behind it. At Turton this applies more to physics than to chemistry. Asked about their future, their reply is often the same: 'I don't want to do physics,' and sometimes chemistry is included.

Could the actual content of the physics and chemistry lessons be one source of discrimination? Both physics and chemistry are almost entirely descriptive for the first year and mainly so in the second year. In physics mathematical work is introduced early, with measuring and calculating, although this is still within a framework of descriptive work. In the third year both become more theoretical, to give some idea of what opting for the subject will mean. For example, both physics and chemistry deal with atomic structure (the work is co-ordinated between the two subjects), but in chemistry it leads into work on bonding, which is an abstract concept, whereas in physics it leads to work on electricity. In my experience boys at this stage not only already know more about electricity than girls, they are also far more interested in it. I suspect that this particular topic may permanently alienate some girls from physics.

Among the less highly achieving girls, the situation by the middle of the third year is even worse. Many of them are quite open about their lack of interest in both physics and chemistry: already they see their future in very traditional terms, such as typing or secretarial work (but only until they get married and start a family). To these girls the physical sciences appear largely irrelevant. I suspect that many of them never consider whether they could 'do' the subject; any attempt is pre-empted by the decision that it is irrelevant.

These attitudes are illustrated by the decisions the students make towards the end of the third year about their future options.[6] From the beginning of the fourth year, students are divided into teaching groups within each subject, to aim for O level, O level/CSE or CSE.[7] In chemistry, currently some 45 per cent of the pupils in O level groups are girls. The corresponding figure for physics is 25 per cent. To put these figures in a broader perspective: the majority of the high-achieving students in the school are choosing to study chemistry in the fourth and fifth year, and this applies almost equally to boys and girls. A similar proportion of the most able boys are studying physics, but fewer of the girls. It is evident from this that there is a significant proportion of scientifically capable girls who study chemistry but not physics, and such a combination of subjects precludes the later choice of many scientific careers.[8]

However, the option choices of the pupils who are average and below-average in science are rather different. A much smaller proportion of this group opt for physics and/or chemistry anyway, and only a small proportion of these are girls: currently around 25 per cent of the CSE chemists are girls, and less than 5 per cent of the CSE physicists are girls.

A contributory reason for this discrepancy between the patterns of choices for high achievers and low achievers may well lie in the perceived career opportunities of these girls. As has already been mentioned, many of the high-achieving girls are strongly motivated towards scientific careers such as medicine, veterinary science, pharmacy or radiography. Clearly most of these girls are not going to expect to find employment locally, whereas many lower-achieving pupils (both boys and girls) do. It is my impression that the employment situation in Bolton does not encourage girls to consider studying science subjects as preparation for scientific jobs. Many boys who leave school at sixteen obtain craft apprenticeships in engineering. (It is also true that the ambition to do this encourages many boys to study physics in the fourth and fifth years.) However, the engineering industry in the Bolton area appears reluctant to offer craft apprenticeships to girls, and many teachers, including careers teachers, are unwilling to suggest this as a possible option for girls when they might subsequently have difficulty obtaining an apprenticeship. I also suspect that the parents of these girls often have very negative attitudes towards their daughters' looking for engineering apprenticeships; perhaps even more than the girls themselves, they do not see this as a 'suitable' occupation. It is true, too, that some male teachers (not only scientists) find the idea of girls applying for jobs in engineering amusing, and sometimes show their amusement in front of the girls concerned, which undermines their confidence.

So far I have dealt with physics and chemistry mainly from the students' point of view: their reactions and behaviour, their subject and career choices. I would now like to return to what goes on in the classroom, particularly in the first three years, and consider the ways in which we, as teachers, might influence (for better or for worse) the progress of girls in physics and chemistry.

I am certain that an important factor operating within schools to influence high-achieving girls for or against the physical sciences is the composition of the teaching staff.[9] I wonder what girls studying science must make of an all-male physics and/or chemistry staff (be this in a mixed school or, even more strikingly, in a girls' school). In contrast, I remember my own days at a girls' school,[10] where science was taught only by women (as was everything else), and it never occurred to me to consider that science wasn't a usual occupation for women – not, that is, until I was at university and found that men outnumbered women on my course by nine to one!

However, there seems to me to be little to suggest that the presence of women teachers will influence the low-achieving girls towards science: such a woman is still a teacher and as such cannot act as a role model for these girls. For them a far more important model would surely be women working in, for example, the engineering industry, and older girls they know going on to craft and technician apprenticeships, with the requisite physical science background.

The sex of the teacher may, however, affect the image of the subject which is presented to pupils. I think that women and men teachers of physics and chemistry have, in some respects, different expectations of their pupils and of the subject itself. One of the ways in which this difference manifests itself is in the issue of the presentation of work. Most teachers would agree that in general girls produce neater work than boys. I find that I am not very tolerant of the often rather untidy work which boys tend to produce. Many men teachers (scientists at least as much as the rest) are more tolerant of poorly presented work, and often undervalue neat work: not because it is neat but, I think, because it is girls' work. Not that I would ever want to defend work that was 'rubbish', but I do feel that girls who are perhaps struggling with the subject, and who write overlong answers through lack of confidence, are sometimes penalised because they have written neatly at length rather than briefly and untidily as many boys would. Is 'neatness' from girls a result of being brought up to try to please people, which is less the case for boys? Is it relevant that girls, up to about age fourteen, have on average superior verbal and linguistic skills to boys? Surely the girls should gain some advantage here? I think they do, and it is most evident when 'free response' writing is required—generally in first and second year and again in the sixth form. Ironically O level and CSE examination questions tend to require short answers, so the girls gain little or no advantage.[11]

I feel that in some respects it is very difficult for men physics and chemistry teachers not to discriminate against the girls in their classes, even when they are aware of the possibility of this and have made a conscious decision not to do so. For example, a great many physics and chemistry textbooks (especially those intended for either younger or less able pupils[12] refer to the student exclusively as 'he' and reinforce the masculinity of the subject through the pictures they use. In these circumstances it is hardly surprising that many male teachers use these same masculine terms uncritically. A few examples will illustrate how the message is conveyed to both the teacher and the pupils.

1. The Nuffield chemistry (1966) O level handbook (intended for teachers) has eleven photographs of students: eight of boys doing some of the most difficult and interesting experiments in the book, one of girls doing a simple experiment, one of boys watching a demonstration and one of boys and girls watching a film. Both the teachers shown are men.

2. McDuell (1976), a modern book for pupils studying CSE chemistry, has a brightly coloured photograph on the front cover of a boy doing an experiment, and a girl standing behind him recording the results.

3. In Chaplin and Keighley (1974), a text for pupils studying physics to CSE, one of the authors is a woman, and some of the questions mention a girl as well as a boy. Nevertheless the student is referred to as 'he' throughout. The book contains sixteen photographs showing scientists at work, but only two show women; in one, a woman horticulturist is standing behind a male colleague, and one photograph features a woman—she is a nurse!

The same masculine language is almost universal in physics and chemistry questions, where an effort is often made to make the subject matter less abstract by referring to students carrying out experiments. However, the questions inevitably continue '. . . he found that . . .', never, in my experience, '. . . she found that . . .'. Many men science teachers seem to find any objection to this usuage trivial or even incomprehensible. They claim that the term 'he' refers equally to men and women—much easier for boys to believe than girls!

A similar problem arises when teachers make reference to examples from everyday life to illustrate some point of theory. In physics, especially, many of the traditional examples (such as references to engines, motor cars or motor bikes) are likely to be outside the experience of the girls. On the other hand a deliberate attempt to make an example suitable for girls, by referring to some domestic situation, can easily become a 'put down'. I am not against the use of domestic examples, but I feel that it is preferable to use an example with which both sexes will be familiar, (and often references to domestic situations will come into this category), rather than to give one illustration for the boys and an alternative one for the girls.

I have found that girls and boys react differently when both sexes are together in a class; in particular I have observed with many classes (first to fifth year) that girls are less willing to answer questions than boys, even when they are equally able to do so.[13] This applies especially to the type of question which occurs quite often in chemistry: 'Can you suggest how we could find out . . .?' Boys will make suggestions straight away, often even before they have considered the question at all, while the girls are frequently unwilling to answer unless it is a factual question to which they are certain of the answer. It is not uncommon in such a situation to find that about half the boys have their hands up to answer the question, but none of the girls. I wonder whether a male teacher would find this as odd as I do? If this happens I never (at this stage) ask a boy to answer the question. To do so would be part of a process of progressively eliminating the girls from chemistry. Yet it would be much easier to ask for an answer from one of the volunteers than to work on and develop the question, perhaps in smaller stages, and to ensure that the girls are involved. The lack of concern which many boys exhibit as to whether they have the correct answer or not often works to their advantage, and hence to the disadvantage of the girls: the girls hesitate, perhaps because they are already conscious that it is not considered 'their subject', and also perhaps because they prefer to offer answers only if they are confident that they are correct.

For the last year, however, I have been teaching one group to whom much of the above discussion does not apply. This is an O level group, where the girls outnumber the boys by two to one (twenty girls, eleven boys),[14] and additionally, of the eight or so most able pupils in the group, only one is a boy. The usual situation is reversed: the girls tend to dominate the group, and I have to work hard to involve most of the boys. But the way the group operates is noticeably different from any other O level group that I have taught: as a class they are quieter, less rowdy in their approach to work. They consider new ideas more carefully before accepting them and the more sloppily executed and presented work of many of the boys stands out because it is not the norm. My experiences with this group make me more determined than ever not to allow the boys to dominate the girls in any class I teach.

Are there any general conclusions to be drawn from my experience? I think that the most difficult problem facing us is the societal one: the pressure on girls, from an early age, to conform to traditional sex-role stereotypes. In this problem I include the unwillingness of industry to offer craft apprenticeships to girls, which is important because the situation in science is far worse for low-achieving girls than for the high achievers. On the more positive side, we can, as teachers, improve the image of physics and chemistry that we project in the classroom, and we can also look for, and demand, less sex bias in published material. Finally we should consider attempting some positive measures to encourage girls to see that physics and chemistry can be relevant to, and useful in, their own lives.[15]

Notes

1 I should like to make it clear that the observations made, and opinions expressed, in this chapter are my own, and do not necessarily coincide with those of my department or of the school as a whole. In addition, I have discussed many of the issues raised with teachers in other schools, and any comments made or implied about the attitudes or behaviour of other teachers do not always refer to my colleagues in Turton.

2 Each academic department in the school does its own setting, and a great many pupils are in A sets for some subjects and C sets for others, though naturally some are in either A or C sets for almost all subjects.

3 Just how early this socialisation to conform to female roles occurs, at any rate with Italian girls, is shown by Belotti (1975).

4 Some evidence for this differential treatment of girls and boys is presented by Newson *et al.* (1978). For a fuller discussion of its effects see Sharpe (1976) and Chapter 4 in the present volume.

5 It is relatively easy to distinguish between the attitudes of these two groups, because at this stage the students are in sets for science.

6 The option choice system we use involves compulsory English language and mathematics for all students. Each student also chooses five subjects from a five-column option grid (which is common to all pupils and includes both academic and vocational courses). Each column offers between seven and nine choices. Popular subjects (e.g. physics, geography) appear in two columns. There are some restrictions, and certain combinations are recommended, but none of these concerns science subjects. At the present time about 98 per cent of third-years choose to study at least one science subject (chemistry, physics or biology).

7 At this stage the division into groups is far from final, and is not a decision about who will take which examination two years later. It does, however, provide a convenient way of distinguishing (in general terms) the higher-achieving scientists, including those who are potential science specialists, from lower achievers.

8 For example, it precludes not only careers directly involving physics

but all branches of engineering, medicine, dentistry, some branches of applied chemistry, and even the study of pure chemistry at some universities.

9 At Turton we have one woman physicist and one woman chemist among eleven physics and chemistry teachers.

10 See *Women and Education* (1978) for a fuller discussion of this.

11 *Cf.* Hardings' analysis (Chapter 14) of sex differences in multiple choice, structured and essay type examination questions. [*Editor*]

12 Many physics and chemistry books aimed at O level students, and almost all those designed for higher levels, are completely asexual.

13 The wider implications of this are discussed by Spender (1978).

14 It is sheer chance that the numbers in the group worked out this way; other parallel groups have an excess of boys.

15 I should like to thank Jill Norris and Judith Summers for reading drafts of this chapter and discussing them with me; and Brian Arnold, my Head of Department, for his help with its preparation.

References

Belotti, E. (1975), *Little Girls*, London: Writers and Readers Publishing Co-operative.

Chaplin, S., and Keighley, J. (1974), *Focus on Physics*, Exeter: Pergamon Press.

Department of Education and Science (1975), *Curricular Differences for Boys and Girls*, Education Survey 21, London: HMSO.

McDuell, B. (1976), *Foundation Chemistry*, Sunbury on Thames: Thomas Nelson.

Newson, J., Newson, E., Richardson D., and Scaife, J. (1978), 'Perspectives in sex role stereotyping', in Chetwynd, J., and Hartnett, O. (eds.), *The Sex Role System*, London: Routledge & Kegan Paul.

Nuffield Chemistry (1966), *The Sample Scheme Stages I and II*, London and Harmondsworth: Longman/Penguin.

Sharpe, S. (1976), *'Just Like a Girl': How Girls Learn to be Women*, Harmondsworth: Penguin.

Spender, D. (1978), 'Don't talk, listen!', *Times Educational Supplement*, 3 November, p. 19.

Women and Education Newsletter (1978), No. 13, 'Girls schools remembered'.

From teachers' letters

I think that the problem is in many ways historical and is paralleled in many other male-dominated professions. Physical scientists have, for many years, been almost exclusively men, or perhaps more correctly the ones to achieve eminence have been. Either consciously or unconsciously, this has influenced the teaching method to maintain the *status quo*. Most physical science textbooks and courses are designed to appeal to what are socially acceptable masculine interests – mechanical things, industrial technology, explosives, etc. Yet logically there is no valid reason why this should be so. For example, I think that one of the best methods of teaching the mechanisms of heat transfer is to relate them to cooking.

What I am really saying is that the problem is of our own creating. We do not offer in schools physical science courses which are suitable for both boys and girls; instead we offer courses designed for boys, which the girls can take if they wish or dare.—*Male physical science teacher, mixed secondary modern school.*

In the early stages, there is in my experience little or no difference between girls and boys. Where boys gain because of greater technical experience, girls gain by being more painstaking and respectful of apparatus, and in theoretical understanding eleven-year-old girls are frequently ahead of the boys, in this as in other subjects.

As the pupils get older two things commonly happen, both of which I think are significant. Firstly, the work in physics and chemistry becomes more mathematical, and secondly, it is more often taught by a man. Boys who are weak at mathematics are made to feel almost that this is a slur on their manhood, certainly they will not 'get anywhere' without mathematics. The same pressures are not placed on girls, the attitude is more resigned than anxious, and the girls feel they have been given permission to give up.

The sex of the teacher is important from the point of view of teacher attitude and expectation, which has been shown by American studies to be strongly related to pupil achievement. A male teacher, as any other, will emphasise automatically those aspects which most interest him. These are likely to be those which interest boys, but not necessarily girls, because of the different social conditioning they have experienced. The girls lose interest, and the teacher expects neither interest nor achievement. A female teacher, on the other hand, not only presents the female view, but knows that girls can do science, because she herself succeeded, and is also living proof to the pupils of this fact. If she is also successful in other female spheres, so much the better for the image.—*Female physics teacher, mixed comprehensive school.*

There are in my experience a number of reasons why girls do not in general choose physical sciences in many schools, and I list them in what seems to me to be roughly the order of effectiveness:

1. *Prejudice* of most headmasters and headmistresses in favour of the 'Arts' subjects, which are wrongly regarded as more 'cultural'. This prejudice is reflected in the allocation of periods in the timetable and in the number of staff engaged. For example, it is regarded as quite wrong by many that an O level candidate should do three sciences for O level but

commendable that a pupil does English, French and, say, German or Latin. Many headmasters regard biology as 'useful in nursing', etc., and hence 'good for girls'. Since relatively few boys do biology (shortage of career opportunities), this sometimes becomes a 'girls' subject' by default. Views are expressed in public about the 'narrowness' of a science, and this affects girls more than boys because the career opportunities for boys more obviously outweigh any imagined cultural disadvantages.

2. *Mixed schools.* Many girls lack the self-assurance which boys of the same age group show when tackling practical work, and some even cultivate the 'helpless female' syndrome to get help from the boys. They end up feeling that physical science is a 'boys' subject' even though the able ones are often better than the boys. Some masters have told me (at ASE meetings) that they deliberately frighten off the girls – especially in the first term of the sixth form – by making the work difficult. Girls who are more conscientious and less confident drop out, and then they revert to the normal standard of work.—*Physics teacher, girls' convent grammar school.*

It is a fact that *most* girls have not the type of mind that faces a problem, nor reasons well from given data – not even my star girls, who got as far as Oxford and/or Cambridge, and to a first and a PhD in one case. Even this girl just could not compare with her boy rivals but she took the subject further and at a university where standards were lower.—*Male physics teacher, mixed grammar school.*

Our system of science education is geared to the male ego. I will give some reasons:

1. Most secondary schools do not standardise their exams to a common mean and deviation. Girls will often get much higher marks in English/French/History, etc., than in science. This will affect boys too but they *don't care* so much. If they like science they'll stick it, but girls may give up because they think they're not 'good' at it. Being 'good' at something matters more to girls.

What can be done? Insist that secondary schools standardise their internal examinations – maybe to a mean of 60, standard deviation 15? *Or*, if the school won't, then the Science teachers can play the game too by converting their raw scores of right and wrong to a scale comparing favourably with other subjects.

2. Science syllabi are more overstretched than Arts syllabi, and there is the tendency for what was A level work in say 1950 to become O level work twenty-five years later. There is the continual danger of presenting *each successive year* concepts that are just at the fringe of pupil understanding. Most boys will say, 'Who cares if I don't really understand? The experiments are fun, and I usually do understand in the end,' i.e. DO, THEN UNDERSTAND, but the girl usually likes to UNDERSTAND AND THEN DO. It must also be remembered that the experienced expert teacher can make fringe concepts understandable, whereas the inexperienced average teacher will often present them in a muddling way. And which sex receives more science teaching from experienced experts? And even if men teach girls there is often not the same conceptual empathy to help put over difficult concepts.

To remedy this we need very careful Piaget-type attention to concept difficulty – gearing it to what the average inexperienced teacher can put over. This will mean pruning the tops of O or A level syllabi, and allowing difficult concepts to become options set against, say History of Science to allow girls to display their strengths. In scientific *work* the female aptitude to take pride in repetitive achievements in what is fully understood has just as much value as male willingness to dare what is partly understood.

Maybe your letter provides an important clue. You enquire 'why boys *do better* than girls in science, and what, if anything, can be done to improve girls' *performance*'. I underline three key words which are sex-biased:

(a) *better*: Girls have less competitive instinct. They enjoy team work. And what does workaday Science need?
(b) *do*: Scientists spend too much time *doing* things. At university we scientists had little free time for the Union, for debates, for learning social communication skills with others. No wonder the arts students gained the top management positions. Our country needs scientists who are able to *be* people as much as *do* science. I might even discourage my daughters from taking up Science because I want them to develop their capacity to *be* during Higher Education more than their capacity to *do*.
(c) *performance*: This word implies criteria. If the criteria are not fair to the qualities which our society has by *nature* and *nurture* developed in girls, then these criteria must change if it wishes girls to perform as well as boys.

Does the UK *really need* more women scientists? If not, then just leave the present discriminatary science well alone, and stop feeling guilty about it.—*Female zoology graduate, ten years' experience science teaching, now remedial reading teacher.*

The main reason for girls' choice in science subjects is their commitment, as women, to an interest in personal relationships. An intense interest and affection for animals (possibly the developing mother instinct?) arouses and sustains interest in biology.

The girls find little in common with their experience in physics and chemistry. They have little interest in quantitative work and, in particular, the mathematical content of experimental work, and science theory.—*Male head of science, mixed comprehensive.*

Social conditioning is, I feel, the greatest factor in preventing girls from studying science. Firstly conditioning in the home from a very early age – in the home, garage, etc.

At school, those girls who still have the confidence to pursue their natural curiosity in science eventually realise the male-dominated world into which a science qualification will lead them.

One has to be something of a pioneer to study science at school. Many times I have encountered amazement from other women when I 'confess' to teaching chemistry. This attitude is felt acutely by many adolescent teenage girls studying science. For a girl to take up such 'masculine' careers as engineering must require some measure of courage.—*Female chemistry teacher, girls' grammar school.*

Biology has never been short of capable and indeed outstanding girls. My answer is that the subject remains descriptive and more susceptible to rote learning than physics and chemistry. Girls remain conned into thinking that such intellectual activity is their forte.

Chemistry has had a few female stars. They, in my opinion, resisted or ignored the social pressure which implies that chemistry is a man's world – aggresive acids, nuclear power, menacing odours and some mental gymnastics needed to cope with the mole. Would it be too easy to suggest that the twelve or thirteen-year-old is more likely to receive a chemistry set for Christmas than is his sister? Is it too obvious an answer that success in chemistry means loss of feminity?

Physics has even fewer girl successes. Here, however, the masculinity of the subject is more strident – electronics, speed, pressure and mathematics.—*Male chemistry teacher, mixed comprehensive school.*

The mothers of our pupils often regard science as an optional thing and will sympathise when their daughters find it difficult, whereas in other subjects they would urge them to work harder. The girls also find fathers and brothers inclined to deride their science. To some extent this is part of a general idea that certain subjects are for girls and others for boys, and the men in their lives rather resent the girls' knowledge. This sexual division was most marked in the mixed school where I taught briefly and is probably one reason why girls in such schools seem to do rather less well in sciences than those from girls' schools. I think that boys are willing to use terms that they do not fully understand and this frightens the girls, who are less inclined to talk like this.—*Female physics teacher, girls' independent school.*

From my experience, boys don't do better than girls in science: it all boils down to the matter of opportunity and encouragement plus the provision of adequate facilities and apparatus.—*Male science teacher, independent prep school.*

It would be insidious to name the schools concerned, but my experience of two grammar schools in North London might be of some interest. One school was a mixed school, the other a single-sex girls' school. Since they were in roughly the same area one might have expected that the intake would come from a similar background and show a similar pattern of achievement. This was not the case. The results were parallel in both schools up to the end of the third year. In the fourth and fifth years the girls in the single-sex school carried on and reached the standard I might have expected from their prior performance in years 1–3. In the mixed classes the girls' performance deteriorated after the age of puberty. A similar pattern was shown by girls in a near-by mixed comprehensive. Apparently pressure was brought to bear by both staff and parents to the effect that, *being girls*, their opportunities would be much more limited in a science field. Unfortunately some did opt to continue biology and at a higher level found their lack of chemistry and physics to be a disadvantage. From many years of teaching chemistry and being female myself I believe the above situation is quite common, especially if the science or school hierarchy is male. Parents and many teachers do not think of physical sciences as 'girls' subjects'.

This is not the only reason why boys do better than girls. I think the peer group, fourteen to sixteen considers that science is somewhat unfeminine *just at a time* when the girls are becoming *aware* of their feminity.

A further possible factor, difficult to prove, may lie in the different type of intellect possessed by boys and girls. Girls tend to be more practical, more easily stimulated by visual and pictorial teaching and are less able to cope with pure abstractions. Perhaps a different approach, teaching the same concepts with more examples of direct application to life, would be more successful.

Women have been tied to the home in the past and have had little opportunity or inclination for 'flights of fancy'. Although this is changing, attitudes still remain and are handed down. Many females become impatient with the abstractions of the physical sciences, mathematics and philosophy, because they cannot see any practical use for them; it may not be that their intellect is different, just their motivation.—*Female chemistry teacher formerly at mixed and single-sex grammar schools, now at college of further education.*

I have recruited 40 per cent of my O level Physics set to be girls. The proportion has steadily increased to this level, and I see no reason why this percentage should not at least be maintained now. I do positively recruit girls and I give a lot of attention to their equality of opportunity although I think that the effect of society itself in its equal expectations is getting through to the present generation of young girls whereas the generation of five years ago was only half getting this message.

Perhaps I should qualify my word 'recruiting'. I do go to talk to all third form upper band physics groups about careers and about how I see the subject of physics in relation to the whole area of careers, emphasising how both boys and girls fit into this whole plan. I think that it is crucial for a committed science teacher to do this, particularly for girls at thirteen years old (and twelve too), in person, and preferably more than once. I even talk to our infant school six and seven-year-olds about electronics and maths – it's a little early yet to know what effect this will have on their university class of degree!

During the O level (and A level) years I give particular attention to relationships, preferring over-familiarity to fear, capitalising as much as possible on convenient similarities and convenient differences between the sexes, such as the greater maturity of the girls with respect to the boys. But, of course, I do have to put up with the girls' off-days and their 'frisky days' and also the needs of the boys for just that little bit of extra firm 'mothering'.

Girls tend to start their fourth year in groups, but gradually I do get them to integrate into mixed groups; nevertheless, girls are more obviously group people and rather better communicators. The loner girl does need very close watching.

As an exponent of a proportion of individualised learning in physics, I do find that this technique is particularly suitable for girls to really switch on when they are in the right mood, which is, of course, for most of the time, but work can be postponed temporarily when the need arises and made up for at a later date.—*Male head of physics, mixed comprehensive school.*

Here is what I do to try to get improvement among girls in the sciences:

1. I try to find out what the class likes and dislikes . . . what science topics they like . . . how they like them taught . . . what homework they prefer . . . how they like to express their ideas and knowledge. This is what I call *class personality*, and it is not easy for me to ascertain.

2. Having discerned their science likes and dislikes, I select topics and teaching techniques *to feed what they like*. For example, in year three (thirteen-plus) colour, music, photography and astronomy are very popular.

3. Suppose I select COLOUR for them. Their first homework would probably be a science essay entitled 'colour'. They explore the library for information and illustrations and very much enjoy composing illustrated essays. This is *literature research*, and girls certainly like it. (The school library has to have plenty of up-to-date books at the appropriate level of difficulty.) So I find *literature research* beneficial for girls.

4. Continuing colour . . . in class, their double period is half practical and half writing-up, with teaching talk interspersed. Their practical would be, for example, making primary coloured lights / mixing lights / complementary colours / white light / Newton's disc / paint colours / the spectrum of sunlight / prisms / the rainbow / stage lighting / colour harmony and contrast / effect of light on coloured fabrics / etc. Now, all these items are selected from the O level syllabus . . . so as well as catering to an interest in year three we are preparing them for O level. This I call *Two for One*, and it keeps down the total amount of academic slogging they have to do (which is a lot!).

5. Continuing colour . . . in their writing up they like to have a *strict routine*: title/apparatus/method/result. In year three they like to emphasise drawings and apparatus. So I emphasise . . . *coloured drawings* and especially those which explain what is happening. This is using what I call the *Chinese motto*: 'a picture is worth a thousand words'. By and large, girls are talented in using drawings to explore and convey science. I call this *Science Art*.

6. Continuing colour . . . I find that the girls like *very clear instructions* on what to do and how to do it. Once they have 'done it right' and 'written it right' they are then relaxed. They like to get their duty done. Once they have done their duty they return to the experiment and start playing with it, exploring new effects, discovering, asking endless questions. So what I have learnt is that they progress according to the motto: 'Duty before Discovery'. (Possibly boys like it the other way round.)

7. Likewise in year three, I teach them MUSIC (O level sound), PHOTOGRAPHY (O level light) and ASTRONOMY (O level gravity). This is *Exploring Useful Topics*.

8. In years four and five I have to abandon the previous approach, i.e. topics . . . and stick to the stark syllabus. This kills some of the interest built up in year three. At this stage their ability in mathematics is important to their enjoyment and progress.

There is also the question of personality . . . at present, science is historically organised to cater to the job needs of boys, i.e. to the personality of boys. Now many science teachers would, I feel, say: 'Nonsense! Science is science. It is neither male nor female.' True indeed. Knowledge is knowledge. Knowledge has no sex. However, personality strongly comes into:

How knowledge is explored.
Why knowledge is explored.
How knowledge is used.
Which knowledge is used.
How knowledge is enjoyed, etc.

When it comes to science, we have to be aware that male personality is not the only personality produced by Evolution. I find that girls enjoy science in ways different from boys. To help girls, there is a need for *textbooks* which appeal to a wider range of personality and interest and usefulness. Although the sexes are 'equal' they are 'different' and it might be that this difference is now being overlooked in the trend for equality. The modern 'personalities' of science which appeal to girls are, as we all know, such as medicine and animal sciences. However, there are certainly hundreds of lesser known skills which would attract girls, e.g. in genetics, crop improvement, brain research, public health, etc. I think there is an urgent need for a special *series of films* (for schools) which would show boys and girls their separate opportunities. These films could be accompanied by colourful *booklets.—Male head of physics, girls' independent school.*

19

Sex typing in schools

JUDITH WHYTE

The great majority of teachers genuinely wish not to discriminate between their pupils on the basis of sex; but it is possible that they do so unconsciously and unintentionally. At present little effort is made to sensitise future teachers to their own unconscious biases or to suggest ways in which girls' under-achievement in crucial areas such as maths and science could be combated. In initial teacher training courses there is not even much attention paid to the large body of theoretical work on sex differences (e.g. Maccoby, 1963; Oakley, 1975). In this chapter I want to mention some of the research which has direct implications for science teachers, and to suggest compensatory and other strategies for change in the classroom.

The process which 'turns' girls 'off' science is complex and starts early (see Chapter 4). Girls' under-achievement in maths and science may be directly related to insufficient experience of mechanical and constructional toys and a lack of early activities at home and in primary school, involving spatial awareness. In addition, children quickly learn which activities and occupations are considerd appropriate to each sex, and a girl's attitude to sciences may well be 'this is a boy's subject at which I am not expected to be any good'. These two factors, the objective lack of background knowledge and the subjective anticipation of failure, can only partially be remedied by science teachers alone. So I will have to expand this discussion to show how structural and organisational factors in the school, together with the attitudes and behaviour of other staff, impinge upon girls' scientific achievements.

The 'hidden curriculum'

As well as learning enshrined in the public or 'open' curriculum a good deal of learning of other kinds goes on in schools. In what has been termed the 'hidden curriculum', pupils receive invisible lessons in sex role acquisition, in how to approach personal relationships and sexual behaviour and in what should be their occupational choices.

The very geography of the school may have its own implications:

> The school has two playgrounds—one for infants (5–6 years) and one for juniors (7–11 years). But the girls never graduate to the junior playground; they stay with the infants. The junior boys are expected to be too boisterous for the younger children, but the girls present no threat. No ball games are allowed in the infants'/junior girls' playground. (Harman, 1978, p. 8)

Wolpe (1977) describes a similar situation in a co-educational comprehensive school. There are two playgrounds, one mixed and one for girls only in which no balls are allowed. Although it was originally intended to have a quiet playground open to both sexes it was found that chaos resulted from the boys rushing from one to another. But the boys monopolise the mixed playground with football, with the result that there are no opportunities for the girls to be physically active during school breaks. As Wolpe says, 'the situation is structured in such a way that they have little alternative but to be onlookers'.

At first sight this has nothing to do with science achievement. But it may be relevant in several ways. Where football is a predominating sports interest in the school, excluding girls may have the effect of making them feel inferior and less important than boys. This obviously links up with their lack of self-confidence, and perhaps with their resulting lack of originality and willingness to experiment (in the truest sense), in the science classroom. There may also be some connection between a child's daily physical experience of three-dimensional space and the development of spatial ability. Thus boys' greater opportunities for rushing about may link up with their greater spatial ability. Moreover the school's informal provision of outdoor and physical activities encourages the assumption that qualities of energy and adventurousness are the monopoly of boys. It makes no allowance for the quiet boy or the energetic girl.

As a counter-measure, playgrounds could be made 'unisex', and schools and teachers could positively encourage girls to play football rather than actively preventing them from doing so. The 'spin-off' effect into the classroom might prove surprising.

In general, as teachers, we should not assume that separation of boys and girls is natural: 'after all, they are different . . .'. Separate lines, roles, duties and punishments exaggerate the differences between the sexes at the expense of what they have in common as learners. Such separation encourages the pupils to think that what boys do and what girls do is different, and may make them more willing to categorise science as one of the things that boys do.

The same point can be made with respect to some seemingly innocuous bits of chivalry. The marked decline in girls' academic performance in the first three years of secondary school coincides with a stage in life when both sexes are confronted with the problem of 'becoming', in all senses of the word, a man or a woman. Members of staff themselves may unconsciously promote conventional 'feminine' role behaviour in the girls they teach. Wolpe reports the case of a teacher who told a girl to 'go to Mr A, charm him, use a lovely voice, say "Thank you, sir," and smile at him when you ask him for the books'. She comments that 'he was coaching the girls and telling them that being pretty, and being nice, and being charming were all essential ingredients to achieving particular ends'. These are all qualities of being feminine. Mr B had 'unwittingly' defined the situation for both boys and girls—the boys would also learn what to expect from the girls (Wolpe, 1977).

Again it is not immediately obvious that this sort of social interaction can actually affect the way a girl tackles a maths problem, or thinks about physics. But the connection may lie in the suggestion that femininity consists of charming a man into doing something for you, rather than learning to do it yourself. As such it leads to female helplessness in science as elsewhere.

The answer is *not* to 'treat boys and girls exactly the same', something many teachers try to do in a sincere attempt to be 'fair'. Sex is a crucially significant element in any social interaction. It is impossible to be unconscious of the sex of our pupils, and a pretence, however well-meaning, of equal treatment only disguises unexamined gut reactions and unspoken expectations. A discussion session among staff about sex differences may prove useful in clarifying for each teacher how her/his behaviour and attitude are operating in the classroom to the disadvantage of some pupils. We know from a study of teacher response to dependent and aggressive characteristics in pupils that 'dependent' girls are preferred to any other combination (Levitin and Chananie, 1972). This may be because dependent girls are exhibiting both pupil-appropriate and sex-appropriate behaviour. Aggressive girls were least liked—

probably because they violated both the professional and personal values of the teachers. The more teachers can increase their awareness of these interactions (through in-service courses, perhaps) the better they can control the classroom environment for the benefit of pupils.

Staffing structures can carry their own messages. If, as often happens in mixed schools, the Head and all the heads of departments (except girls' crafts) are male, and the deputies are female, girls and boys will learn that men lead and women support them (see Davies and Meighan, 1975). The most significant aspect of such 'hidden learning' is that it operates below the level of full consciousness.

Textbooks and teaching materials also carry hidden messages. The sex-stereotyping of children's books, including both primary and secondary science texts, has been thoroughly documented (see Chapters 4 and 18). In the long run it may be possible to find more suitable replacements, but meanwhile many books we may not entirely approve of will continue to be used, and needed, in school. In the short term teachers can adapt their use of existing textbooks, use home-made materials, or just comment reassuringly on the sex stereotypes they come across.[1]

Teacher behaviour

Another source of hidden learning may be the behaviour of teachers in the classroom. Davies and Meighan (1975) report that in the urban comprehensives they studied 'The teachers almost with one accord wanted to encourage absolute parity between boys and girls in timetabling, in curriculum and career opportunities, and even games. Surprisingly however they were then willing to make quite sweeping generalizations about differences between boys and girls both in terms of ability and behaviour.' One of the areas in which differences between the sexes are commonly believed to occur is science. If educators tend to believe that boys are naturally better at and more interested in scientific, mathematical and technical subjects, and that girls are more subjective and emotional, less curious and adventurous, more likely to be interested in literature or domestic subjects, this may become a self-fulfilling prophecy.

Palardy (1969) found that teachers' preconceptions about girls and boys as proficient readers seemed to constitute a self-fulfilling prophecy. The boys did better when they were *not* expected to do worse than the girls than when they were. Rowell (1971) reports a similar effect for science—when science teachers expected boys to do better than girls the sex differences were larger than in classes where science teachers had more egalitarian expectations.

Davies and Meighan (1975) found that, in response to a questionnaire, fifth form girls showed no preference for male or female staff, but made many comments about the differential treatment they themselves received from staff. Teachers were said to be harsher with boys, but to give them more attention or friendliness, while girls were treated as 'helpless' or 'stupid'.

This perception is confirmed by research findings which show that boys tend to get more attention, both positive and negative, from teachers than do girls (Meyer and Thompson, 1963; Brophy and Good, 1974; Sears and Feldman, 1974). The implication for pupils is that boys' behaviour is worthy of teacher attention, and girls' is not. This is echoed in Davies and Meighan's study, where the teachers saw girls as more 'conscientious', precise, better at written work, but 'devious', 'insidious', 'insolent' and 'resentful'. Boys were thought to be more logical, enthusiastic, quicker to grasp new concepts; but their faults were described in Enid Blytonish terms: 'prank-playing', mischievous, naughty, but always 'owning up'. In response to a forced-choice question, 72 per cent of the teachers (both male and female) said they would prefer to teach boys. As the authors observe, 'Girls' complaints about receiving less attention could well be justified if teachers do reveal their appreciation of the boys' dynamic personality characteristics' (Davies and Meighan, 1975).

Perhaps the most vital task for teachers to undertake is to 'de-stereotype' their perceptions of their pupils, and incidentally of themselves, their colleagues and friends. In work with teachers in training I have found the 'de-stereotyping exercises' developed by a team of educationalists at Boston University very useful (Nickerson *et al.*, 1975). A typical game provides adult groups with information about a person working for a bank, and participants have to make guesses, 'fantasise' about the person's job responsibilities. Some groups receive a data sheet headed D. A. Johnson, others an identical sheet headed Denise A. Johnson. Follow-up discussion centres on discrepancies in the profiles of males and females (for example if Johnson was supposed to be a male he usually has more people working under him than if she was a female!) and why these arose.

De-stereotyping in the classroom

Many teachers in the English educational system have a strong commitment to a *laissez-faire* approach in seemingly insignificant matters such as classroom seating arrangements or more important ones like career choice. But the teacher who understands the educational harm that can be done when things are left to happen 'naturally' will become an 'interventionist'.

Leave pupils to choose their own groups in a science class, and the chances are boys will sit with boys, girls with girls. This sex division then firmly stamps the outcome of many of the learning experiences that follow. I observed one biology lesson, by an otherwise excellent teacher, in which the classroom was 'accidentally' divided according to sex with two 'girls' tables' and three 'boys' tables'. Discipline was good; the girls were on the whole silent and attentive; the boys occasionally talked among themselves. The teacher frequently addressed remarks to the boys—'Sit up, Michael!' 'What's the joke, Alan?'—generally as a means of containing any potential disruptions or distractions. No such comments were made in the direction of the girls' tables. Throughout the lesson he stood between two of the boys' tables, at the opposite side of the classroom from where the girls sat. After the exposition, when group work started, I asked about the sex divisions. The teacher assured me that it was 'purely accidental', didn't happen in every class, and in any case was 'their own choice'. He added that he always found the girls to be very efficient and industrious, usually getting on with work on their own and sometimes doing better than the boys. It is impossible to say how much his confident expectation of this behaviour from the girls, and his resulting tendency to interact much less with them (e.g. he more than once 'missed' seeing a girl put up her hand), actually produced the 'quiet, efficient' behaviour. At any rate, the girls participated very little in the teacher-directed part of the lesson, and their acceptance of a less active role was taken for granted. I was reminded of the relationship women used to have to scientific studies in the nineteenth-century universities, 'sitting in on' lectures directed at the male audience of students.

So if science teachers want to get the best possible work from girls they should not assume that when the girls are quiet, docile or seem content with an observer role that all is going well. Quite the reverse. Girls may benefit from being pushed into more 'active' 'performer' roles in science and technical subjects, where manual and spatial skills are often important. Girls' silence should not always be taken for understanding: they may be afraid to ask questions or display ignorance. Or, worse, may quite happily confess their ignorance as a sign of their emerging 'femininity'.

If science teachers are simply aware that, because of its masculine image, girls may need all sorts of positive encouragement to enjoy science, they can begin to prevent female pupils dropping out. At the beginning of a new term with a new class an overt discussion of attitudes to girls and science might be worth while. The teacher should make it clear not merely that s/he thinks girls can 'do just as well as the boys' (which always carries the unspoken implication that other teachers do *not*) but that s/he is actively interested in girls' scientific development, that s/he wants them to do well, and is prepared to give girls active encouragement, because it is appreciated that, especially in physics and maths, lots of people, including parents, do not expect girls to do well. Girls and boys should be allowed to spell out their doubts about this viewpoint, so that the teacher can try to deal rationally and fairly with all their points. It may seem quite irrelevant to a physics teacher with a heavy load of work in the year ahead to spend thirty minutes talking about, for instance, the undervaluing of women's scientific achievements in the past, or discrimination against girls in the engineering industry. But ultimate effects in terms of pupils' enthusiasm and commitment to the class can be immeasurable.

Science teachers might also attempt to talk more to the girls even though they don't seem to require it. This could uncover problems they have been unwilling to voice or errors which would have gone unnoticed. Boys should be rewarded (with praise and encouragement) for co-operative behaviour, or when they have been working silently alone, 'getting on with it', and not (with teacher's attention) for disruptiveness. If nevertheless more teacher time is needed simply to keep the boys under control the teacher who is aware of what s/he is doing can later on plan to compensate the girls with extra attention when there is an opportunity. It should not be assumed that girls will always need help with complicated or expensive equipment. All pupils should be introduced to the apparatus they will require early on in such a way that they can use it independently and competently.

An open and honest willingness to examine one's own stereotypes and an alertness to common but damaging assumptions about what is appropriate to one sex or the other are ways of challenging and altering these assumptions.

Facilities

As well as examining personal preconceptions, science teachers might do well to give critical consideration to some of the *structural* assumptions on which secondary school provision is currently based. For example, if there is a shortage of science accommodation, why should it be assumed that *boys* have a prior claim to the use of scarce laboratory facilities? Eileen Byrne, in a study of resource allocation to secondary schools in three LEAs, in all of which there was overcrowding and therefore rationing of accommodation, found that 'it was by no means uncommon practice to give laboratory preference to boys (taking physics), teaching biology to girls in classrooms' (Byrne, 1974, pp. 38–9). Byrne also quotes a national study by science teachers (Science Masters' Association *et al.*, 1960), indicating greater deficiencies of accommodation in girls' schools, and a higher level of provision for handicrafts than for homecrafts, overall. Consistent evidence appeared that considerably more money was allocated to boys' schools than to girls' schools of equal standing. This economic factor alone offers a partial explanation for the current under-representation of females in scientific and technical occupations.

Nor has co-education necessarily redressed the balance. Byrne also observed that, even when laboratories were freely available in theory to both sexes, Heads created quite separate and differently biased courses and options for boys and girls respectively at thirteen-plus and fourteen-plus, tending to 'channel' girls to biology and boys to physics. If the option system requires children to choose a 'package', then putting office skills with biology, or woodwork with physics, will tend to steer girls and boys into customary sex-linked options.

If pupils are to have a realistic choice of options in the first place, it is essential that they have experience of as wide a range of subjects as is possible before any choice is made. The crafts department will quite rightly be unwilling to accept in third-year woodwork a girl who has no experience in the subject. Some schools have solved the problem by providing rotating timetables so that all pupils spend some time in the first two or three years working at woodwork, metalwork, technical drawing, engineering, sewing, cookery etc. However, it may be that staffing and other resources are linked to option systems in such a way that a sudden demand from girls—or boys—for non-traditional options *could not be met.* The National Council for Civil Liberties' booklet on sex discrimination in schools

(Harman, 1978, p. 25) recommends that we 'train teachers in timetabling and the allocation of scarce resources so that they can do this in such a way as to avoid the division of children according to their sex'. This might entail arranging options so that all pupils had to have a grounding in two or even three science subjects (for instance, by providing an integrated science course right up to O level). One science subject (particularly if it is biology) is inadequate for most technical occupations and scientific careers.

Careers advice

Vocational education and course career guidance usually varies by sex. A useful booklet published by NCCL notes that 'Careers education is insufficient, too late, and acts merely to reinforce traditional views' (Harman, 1978, p. 14). Harman recommends that careers teachers should have special in-service training to make them aware of their responsibility for widening opportunities for girls; time should be allocated for careers education before children make their subject choices, and careers teaching should include giving girls a realistic understanding of how many years of their lives are likely to be spent in employment.

It is at the option-choosing stage that many teachers, perhaps with good intentions, seem to discourage girls from physics or maths, perhaps because they regard it as unusual or unsuitable for a girl to take all three sciences plus maths; even an indication of surprise at her decision may be enough to put a girl off—she may feel that her efforts are not likely to be taken seriously by the teacher, and her self-confidence may not be high enough to withstand such unspoken pressures.

Yet not enough girls have the necessary qualifications for entering medicine, for instance, and even traditionally female jobs may require a more thorough grounding in the sciences than the majority of girls receive. Trainee nurses would benefit from having qualifications in chemistry as well as biology, and technical skills are increasingly at a premium in modern 'technological' hospitals. If we want to promote primary school science then more primary school teachers must be scientifically literate, which means that more schoolgirls must obtain maths and science qualifications.

Some careers teachers, science and technical teachers among them, may overemphasise the difficulty for a girl of crashing into a male preserve like engineering (because of the isolation and discrimination she will encounter), and so discourage her from gaining the relevant qualifications. It would be more 'fair' to forge links with local engineering employers and, for example, persuade them that it will be worthwhile taking on a cohort of girl trainees in two years' time, thereby seeing the task of careers guidance as being to create more opportunities for girls rather than to reinforce traditional job stereotyping and the sexual division of labour.

Crafts

One of the greatest areas of sex differentiation is in crafts, and this persists in spite of the Sex Discrimination Act.[2] Some schools adopt a policy in relation to crafts which is only superfically egalitarian: 'All boys do cookery; some of them are going to be chefs' carries the unspoken implication that girls will never put their culinary skills to an occupational use, but are being prepared for an inevitable role as housewife. The notion of equal opportunity can end up being subverted if there are separate sex classes (boys do 'serious' metalwork, girls make jewellery), or by unthinking traditionalist attitudes on the part of craft and other teachers. 'The girls are very obedient, they always do exactly as they are told,' said a craft teacher in a television programme, 'but the boys, now, they try to think round a problem, kind of short-circuit it.'

It tends to be taken for granted that large numbers of girls will study something called 'domestic science'. If an element of such courses is genuinely scientific, why not introduce material supposed to be of more interest to girls (the chemistry of cooking; domestic technology) into existing science syllabuses (and not just at the lower levels)? By countering the heavy masculine bias which is still a feature of many science syllabuses, more girls might be encouraged to study science. However, it would be counter-productive to use a patronising 'Here's something for the girls' approach when teaching science to both sexes. Instead we should expect boys as well as girls to be interested in home-based technology, in cosmetics (the teenage pimple industry?), in child-care and nutrition.

And what about the remainder of the content of girls' crafts? 'Housecraft' and 'office skills' are both subjects which have an obvious direct application to occupations, yet the comparison with 'boys' subjects' like technical drawing or engineering is not exact. The boys' subjects develop skills which transfer to other school subjects (especially science), while the girls' have transfer value only to the housewife role or relatively low-paid office work (Byrne, 1979). I know of more than one reasonably competent girl proud of leaving school with up to eight CSE's in subjects like child care, cookery, dress, family and community studies, food and nutrition, homecraft, needlecraft, textiles and dress, pottery or art, who despite all her hard work possesses no real qualifications as far as the job market is concerned. What is needed is a new concept of homecrafts, taught to all pupils, which will provide both sexes with the necessary skills to cook and clean for themselves and care for their children efficiently and, where possible, creatively. This should be coupled with a modification of science courses to make them more attractive to girls—and boys—by including material on the technical applications and social aspects of science.

The Sex Discrimination Act

Lastly, and on a different note, for teachers who must operate in a school environment which is indifferent if not hostile to the aim of eliminating sex sterotyping, what recourse is possible within the law?

A full guide to the responsibilities of LEAs and schools under the legislation appears in the NCCL guide (Harman, 1978). In brief, a complaint must usually be made by a parent on the pupil's behalf. Teachers may help by enlisting support from other members of staff, or the parent–teacher association, or by bringing the matter to the attention of the head teacher. Where these steps are unsuccessful, the local branch of one of the teaching unions may help, or a sympathetic school governor or manager. According to the NCCL, test cases are needed 'not only to remedy individual complaints and clarify the law, but also to remind education authorities firmly that the law on sex discrimination does apply to them' (Harman, 1978, p. 25).

The discrepancy between official pronouncements about equal opportunity in education and actual practice is partly accounted for by the subtlety of some of the processes which produce inequality. Full public discussion, as well as some of the developments suggested above for professional teacher training, might elucidate these processes and help towards their removal.

Notes

1 For teachers interested in developing their own materials, or finding non-sexist videotapes, films, books, pamphlets and other resources, the most comprehensives source is probably *Non-sexist Teaching Materials and Approaches*, compiled and edited by Bob White and available from New

Childhood Press, c/o Photography Workshop, 152 Upper Street, London N.1.

2 At a Croydon secondary school boys do woodwork, metalwork and design technology, and girls do needlework and home economics. The school is being taken to court under the Sex Discrimination Act by a girl who wants to do metalwork, woodwork and design technology (Harman, 1978, p. 18).

References

Brophy, J. E., and Good, T. L. (1974), *Teacher–Student Relationships*, Holt Rinehart & Winston.

Byrne, E. M. (1974), *Planning and Educational Inequality: a Study of the Rationale of Resource Allocation*, London: NFER.

Byrne, E. M. (1979), *Women and Education*, Tavistock.

Davies, L., and Meighan, R. (1975), 'A review of schooling and sex roles, with particular reference to the experience of girls in secondary schools', *Educational Review*, Vol. 27, No. 3, 165–78.

Harman, H. (1978), *Sex Discrimination in School: How to Fight it*, London: National Council for Civil Liberties.

Levitin, T. E., and Chananie, J. D. (1972) 'Responses of female primary school teachers to sex-typed behaviours in male and female children', *Child Development*, 43, 1309–16.

Maccoby, E. E. (1963), 'Woman's intellect', in *The Potential of Women*, ed. Faber, S. M., and Wilson, R. H. L., McGraw-Hill.

Meyer, W. J., and Thompson, G. G. (1963), 'Teacher-interaction with boys as contrasted with girls', in Kuhlens, R. G., and Thompson, G. G. (eds.), *Psychological Studies of Human Development*, New York: Appleton-Century-Crofts.

Nickerson, E. T., Dorn, R., Gun, B., Nash, A. S., Speizer, J., and Wasserman, M. K. (1975), *Intervention Strategies for changing Sex Role Sterotypes: a Procedural Guide*, Dubuque, Iowa: Kendall Hunt.

Oakley, A. (1975), *Sex, Gender and Society*, London: Temple Smith.

Palardy, J. M. (1969), 'What teachers believe—what children achieve', *Elementary School Journal*, 69, 370.

Rowell, J. A. (1971), 'Sex differences in achievement in science and the expectations of teachers', *Australian Journal of Education*, 15, 16.

Science Masters' Association, Association of Women Science Teachers, NUT and Joint Four (1960), *Provision and Maintenance of Laboratories in Grammar Schools*, John Murray.

Sears, R. S., and Feldman, D. H. (1974), 'Teacher interactions with boys and with girls', in Stacey, J., Béreaud, S., and Daniels, J. (ed.), *And Jill Came Tumbling After: Sexism in American Education*, Dell Publishing Co.

Wolpe, A. M. (1977), 'Sexuality and gender roles in a secondary school', in *Some Processes in Sexist Education*, Women's Research and Resources Centre Publications: Explorations in Feminism, No. 1.

CONCLUSION

20
Retrieving the missing half

ALISON KELLY

Half the population of the country—the female half—is missing out on science education. Most girls are receiving only a rudimentary education in physical science; most girl school leavers and adult women will readily declare that they are scientifically ignorant. This book has examined the science education of girls from a variety of standpoints. In conclusion I will try to use the ideas and research results that have been presented here to suggest what practical steps might be taken to increase girls' involvement in physical science. Some of these ideas are now being tried out in the Girls Into Science and Technology project in Manchester (Smail *et al.*, 1980). Biological science is not considered in this chapter, since, as the statistics in Chapter 1 show, girls do not under-achieve in biology.

Most of the practical steps revolve round the secondary school, and specifically the science lessons. Despite the fashionable view that 'schools make no difference' there is considerable evidence that school factors *do* affect subject choice. In a survey of Yorkshire comprehensive schools Harris (1979) found wide variations between schools in the proportion of girls studying physical science. The figures ranged from 2 per cent of fourth-year girls studying physics in one school to 66 per cent in another, with similar variations in chemistry. Such large differences between schools are certainly not biological in origin, and are unlikely to be entirely due to socio-economic or subcultural variations. The conditions within individual schools seem to exert a powerful influence on girls' aspirations and achievements.

These school influences operate at various levels. Some can be tackled by an individual teacher within her or his own classroom, while others demand concerted action by the whole science department or the whole school if they are to be altered. Tradition is also important, and it may take several years to build up the feeling within a school that girls can and should do science. A small, committed group of girls that can 'snowball' is often a great help in establishing this feeling. However, school influences cannot be divorced from out-of-school influences and many intervention strategies will only be effective if they spill over from the school into a more extensive campaign.

Inside the classroom

One of the most pervasive educational fallacies is that equality means identity. Identity of outcomes, perhaps, but certainly not identity of treatment. The teacher who says, 'I don't discriminate against my girls; I treat them just like the boys,' is ignoring the fact that we live in a sexually divided society. By the time they enter secondary school girls and boys are well established in their gender identity, and well aware that males and females typically behave in different ways. They have, on average, different background experiences and different personality traits. Whether these differences are biological or social in origin is largely irrelevant to the classroom teacher. But it is important that the teacher should recognise and respond to the differences which exist; it is equally important that s/he should beware of imaginary sex differences and be sensitive to individual variations amongst the pupils.

One very important way in which girls and boys differ when they begin secondary school science is in their prior exposure to scientific situations. Girls and boys typically play with different sorts of toys and have different hobbies and household tasks. Boys' mechanical and electrical toys will develop physics-related skills, while chemistry sets will give them familiarity with chemical equipment; girls more often lack this background. In addition, boys' toys are probably more effective than girls' toys in fostering spatial skills. Teachers should constantly remind themselves that pupils do not come fresh to science but have differing degrees of relevant experience, and gear their teaching to the least experienced. It is all too easy to set up feedback loops so that pupils who begin with greater knowledge and experience in science can pull further and further ahead. One of the first tasks of the science teacher should be to break these feedback loops and ensure that girls start the formal study of science on an equal footing with boys. Rather than just assuming that they are more timid and less interested in science, teachers should provide reassurance and encouragement when girls seem diffident. And above all never assume that the pupils have mended a bicycle or looked at a car engine—the girls probably haven't.

Girls and boys also seem to differ in their approach to work. Although the evidence for this is mainly anecdotal, it is sufficiently consistent to be worthy of consideration. Girls are generally considered to be neater, quieter, more conscientious, more tolerant of routine, more interested in people and better at teamwork. Boys on the other hand are supposed to be more erratic, more likely to produce original or innovative ideas, and more interested in inanimate objects.

If these sex differences are confirmed in research studies, then one way to encourage girls in science may be to place more emphasis on the things they do well. Girls, on average, have greater verbal skills than boys, and they frequently perform better in written work. But this aspect of science tends to be devalued by comparison with mathematical and theoretical approaches (although the ability to write a coherent research report and to draw together strands of evidence from several sources is clearly crucial for a successful scientist). The importance of neat recording and presentation of results in science is often underplayed at school. In addition, teachers often behave as though erraticness, originality and interst in things are more useful qualities for success in science than conscientiousness, tolerance of routine and interest in people; but this is not necessarily so. Much scientific work consists of routine measurement and observation which must be done carefully and conscientiously. Although the subject matter of science often appears to have little to do with people, it is arguable that this impersonality is not intrinsic (see below); and the practice of science is frequently person-orentated, involving teamwork and close co-operation with others. If this were more obvious in school science, then girls might well develop more positive attitudes towards the subject.

This is not to suggest that there should be one science course catering for girls' strengths and another catering for boys' strengths. On the contrary, I would argue that particular efforts should be made to compensate for any weaknesses in pupils' skills. But strengths and weaknesses should be more evenly apportioned between the sexes, with a re-evaluation of the importance for science of the things girls do well, and a recognition that there is more than one way to approach scientific problems.

Girls are also supposed to be less self-confident than boys, and more reliant on reassurance from the teacher that they are doing the correct thing. Many teachers report that girls dislike open-ended, discovery methods and prefer to know where they are going and see the point of the experiment from the start (see Chapter 18). This may have led to initial reports that girls do less well than boys on 'Nuffield' syllabuses. In fact Harding (Chapter 14) found no overall difference between the sexes on Nuffield O levels—possibly because, as Galton (Chapter 13) suggests, few teachers actually employ the open-ended approach. Directive teaching is often characterised, in a pejerative way, as 'spoon-feeding'. But perhaps this merely serves to denigrate girls' skills. Discovery methods could equally well be characterised as 'playing silly games' as the pupils try to guess at things that their teachers already know. Less emphasis on such 'discovery' might be a step towards more efficient, adult ways of learning.

If girls like to feel reassured by their teachers this may have implications for teaching styles. Whereas boys frequently respond well to a bantering tone or to heavy sarcasm, girls are more likely to be wounded by this hectoring approach. They respond better to praise and encouragement. Disorganised and undisciplined classes seem to be particularly unpopular with girls. Teachers should not assume that methods which are successful with boys will automatically be successful with girls, but should be sensitive to sex differences in inter-personal relationships. Competitive, individualistic work may suit boys, but girls seem to prefer a more friendly, relaxed atmosphere. The social context is important to them. Girls are often reluctant to be separated from their best friends, or to be alone in a class of boys. Fox (1976a) describes an American project which successfully encouraged able girls in mathematics by emphasising the social aspects of the class and stressing co-operation rather than competition. If widely adopted, this approach might lead to substantial improvements in girls' science achievement.

Galton (Chapter 13) suggested that teacher-dominated classrooms are unpopular with girls. The 'personal experience' section of this volume gives several hints as to why this might be. Many girls complain of being either chivalrously helped or chauvinistically ignored by their male colleagues and teachers. In controlling the boys in a class the teacher may unthinkingly use remarks which are offensive or belittling to the girls—remarks such as 'stop giggling like little girls/chattering like old women'—and do nothing to prevent the boys insulting and embarrassing the girls. Equally common is the teacher who does not expect girls to be as good at science as boys, and whose expectations become a self-fulfilling prophecy (Rowell, 1971). If the class is expected to be mainly boys by the fourth year, the teacher may gear her or his efforts towards the boys in the early years of secondary school. Teaching may be mainly directed at the boys, with girls 'listening in' (Chapter 19). If girls are not expected to know the answers to questions teachers may not be surprised when they don't put their hands up, or may try to save them embarrassment by cutting short their time to reply (Chapter 18). Encouragement and information about careers in science may be given more readily to boys than to girls because it seems more relevant to them (Chapter 17). So girls are gradually excluded from science lessons and react by dropping out.

The answer here is for the teacher to monitor her or his own behaviour very carefully; being conscious of the ways in which differentiation takes place is the first step towards eliminating it. Some teachers may still find it necessary, for discipline reasons, to treat boys and girls differently in class. But if they are aware of the discrepancies they can make efforts to compensate in other directions. For example, if they find they are spending more time with boys than with girls during lessons, they can make a point of being available to help the girls after class.

But this sort of compensation need not be necessary. As many of the teachers' letters (Chapter 18) make clear, it *is* possible to involve girls in science lessons by gearing the work to their interests and aptitudes. Several correspondents suggest that girls are interested in the practical applications of science and prefer visual and pictorial topics to the more abstract and theoretical work. This ties in with Dave Ebbutt's suggestion (Chapter 15) that what distinguishes 'girls' science' is that it is product-orientated, particularly if the product is decorative or useful. Perhaps theoretical work could be made more meaningful and attractive to girls by carefully pointing out its practical applications. Other correspondents suggest that it is the observational and descriptive aspects of science which appeal to girls, and there are several imaginative examples of this approach to traditional science topics in Chapter 18.

Another aspect of the science lessons which could be altered to encourage girls is the examples and applications used to illustrate theoretical points. At present these focus predominantly on topics (such as guns and explosives, football, motor-bikes and industrial processes) which interest boys more than girls. But it would be easy to incorporate examples from cookery, sewing, health and household maintenance which would appeal to girls. For example, oxidation could be taught with reference to food preservation and heat transfer with reference to heating a home. Simple machines could as easily be tin openers and egg whisks as cranes and car jacks. Again, this is not meant to suggest that different examples should be provided for boys and for girls. The suggestion is that the present bias towards boys in science lessons should be reduced by the introduction, for both sexes, of material with feminine connotations. By building upon girls' existing interests it may be possible gradually to widen these interests so that they do not reject the more traditionally masculine aspects of science. This is particularly crucial in the third year, just before options are chosen. If science seems irrelevant or uninteresting to girls at this stage they will probably drop it irrevocably; boys may be better motivated by career considerations to ignore short-term difficulties or disillusionments and continue with science.

Science is generally seen as being concerned with 'things', whereas girls are typically more interested in 'people' and may be discouraged from physical science by its impersonal and mechanistic approach. Science problems are frequently set in abstract and passive terms (e.g. a block is pulled along a smooth surface at a constant velocity of . . .) without any attempt to link them to situations which are meaningful for the pupils. But science is not necessarily impersonal. It affects the whole range of human experience and human needs, and this should be made explicit in science education through the use of more person-orientated examples. When 'real life' examples are used they generally focus on the industrial and military aspects of science, but this too can be seen as a distortion of a subject which affects every part of our lives.[1] Project Physics (Rutherford *et al.*, 1975) is an example of a science syllabus designed to be taught 'in a humanistic way' by developing links with music, history and literature. One of the aims of this project was to encourage pupils who usually avoid physics (including girls, but also blacks and pupils from lower socio-economic groups) and first evaluations suggest that it has been successful in this aim.

A more humanistic sort of science might also improve girls' attitudes to the social implications of science. Although it is too simple to say that girls are more interested than boys in the social implications, there are some differences of emphasis between the sexes. Ormerod (Chapter 7) found that girls were more likely to choose physical science if they had favourable views on its aesthetic/humanitarian aspects. Discussion of the social implications of science and the way science has transformed the world has very low priority in most science courses at present. But some knowledge of the consequences of scientific advance is probably more valuable for pupils who will drop science at the end of the third year than a half understood smattering of facts and theories. If it also increased the proportion of girls studying physical science beyond third year, so much the better.

The introduction of discussion on the social implications of science might answer another criticism commonly voiced by girls: that there is no place in it for imagination (Hutchings, 1967). This is a damning indictment of a subject which depends critically on imaginative leaps and combinations. But it seems to refer to the 'right or wrong' character of a lot of school science, and the feeling children have that they are jumping through hoops on a carefully laid out circuit.

If more discussion is to be brought into science classes, then something has to be left out. The obvious candidate is some of the more conceptually difficult material. Physical science is frequently considered more difficult than other subjects both in terms of the average marks pupils get and in terms of the concepts involved (see Chapters 7, 9 and 18). This difficulty may have more serious consequences for girls than for boys. Boys are often sufficiently well motivated and self-confident to continue with science in the face of difficulties. But several teachers suggest (Chapter 18) that girls are more conscientious in their studies and more worried by their failure to grasp the concepts. Coupled with lesser self-confidence that they will understand eventually, this may lead girls to dislike and abandon science in favour of easier courses. Alternatively, lack of self-confidence may lead borderline candidates to attempt CSE examinations instead of O levels (the new GCSE (sixteen-plus) examination may have particular advantages for girls). Ormerod's work (Chapter 7) suggests that improving the marks which pupils get in science (which can be done fairly easily within a school) will have little effect unless the level of conceptual difficulty is also reduced. Analyses of the Nuffield O level courses (Ingle and Shayer, 1971; Shayer, 1972) have shown that much of the work in the early years of these courses is beyond the grasp of most of the pupils. It is this which needs to be reformed. Conceptual difficulty is often associated with a mathematical treatment of topics, and it may be this association which leads girls to dislike the mathematical aspects of science (since they do not, on average, do too badly in mathematics itself). Individual teachers have considerable control over what is taught in the first few years of secondary schools and by conscious efforts may well be able to reduce the conceptual level of their material.

Science in schools

There is, however, a limit to what an individual teacher can achieve. Textbooks, syllabuses and examinations impose severe constraints. They *can* be altered, but group effort is more effective than individual initiative.

Many of the problems of girls in science can be seen as stemming from science's masculine image. Girls often avoid 'boys' subjects', particularly around the time of puberty when they are concerned with establishing their femininity. Teachers and parents frequently have stereotyped ideas of what is suitable and interesting for each sex, and so may not encourage girls to persevere with science. Most of the intervention strategies suggested in this chapter are directed towards changing this image or compensating for its past effects on girls' interests and aptitudes.

One way in which it is perpetuated is through science textbooks and teaching materials which refer far more often to males than to females (see Chapters 4 and 18). When girls do appear it is usually either to ask questions and look puzzled, to watch admiringly or to take notes! This conveys to both sexes the message that girls have only a minor part to play in science. Even the continuous use of the pronoun 'he', although it may be intended to refer to boys and girls equally, is likely to make girls feel excluded. Individual teachers generally have to use the books and examples that are provided, but groups of teachers can get together to put pressure on publishers and public bodies. This approach has met with some success in the United States (Federbush, 1974), where a leading publisher has issued its own set of guidelines for non-sexist writing (McGraw-Hill, 1974, reprinted by Children's Rights Workshop, 1976). Another possibility is for teachers to make their own non-sexist material. Given the amount of work involved, this too is probably best done in groups. Group work has the added advantage of encouraging the exchange of ideas, problems and insights as well as combating the isolation which teachers who are trying to work in a non-sexist way often feel.

In examinations the masculine image of science can have serious consequences. Milton (1958) and Graf and Riddell (1972) both found that girls did better on mathematics questions which referred to specifically feminine activities (such as dressmaking or cooking) than on questions which were mathematically identical but framed in masculine or apparently neutral terms. Examples with feminine connotations are rare in science examinations, and examination boards should be more conscious that this may be disadvantaging girls.

Examination boards should also be made aware of Jan Harding's results on different types of questions (Chapter 14). She showed that the present trend towards multiple-choice questions and away from essay-type questions may be disadvantageous to girls. To be fair to both sexes in examinations there should be a balance of different sorts of questions, with perhaps an emphasis on structured questions. There is no justification, in terms of the qualities needed by a successful scientist, for preferring a clear-cut, right-or-wrong multiple-choice answer to the qualifications and ramifications which can be explored in an essay. This seems to be an example of undervaluing the things girls are good at. However, multiple-choice questions will continue to be used for the forseeable future and teachers should consider giving girls extra training and practice in answering them so as to minimise any disadvantage.

Examination syllabuses certainly limit the scope of the innovations that can be implemented within a school. But the impression that examination syllabuses are immutable and impervious to teacher opinion is of course an illusion. Teachers can make their voice heard via subject committees, either by writing to them directly or by attending teacher meetings. The committees welcome new suggestions, and changes do come. For example, the JMB is currently introducing a new O level chemistry syllabus which includes the industrial, social and economic implications of chemistry and omits some conceptually advanced topics such as gas law and electrolysis calculations. Despite considerable opposition an equivalent A level is already in operation. CSE boards are even more receptive to teachers' comments than O level boards, and Mode 3 syllabuses give energetic science departments the chance to try out their ideas. With new syllabuses currently being devised for the GCSE examination this seems an appropriate time for concerned teachers to suggest reforms in syllabus content which may benefit girls.

A dramatic change, which could be considered as a long-term aim, would be to abolish the distinction between natural science and domestic science. Basic scientific concepts could be taught to both sexes with the material currently included in domestic science syllabuses. As Judith Whyte (Chapter 19) suggests, this would undoubtedly give science a more feminine image and render it more attractive to girls. The concepts, theories and experimental testing of science need not change; but the applications and examples could be related to girls' interest by including domestic topics. This would have to be done sensitively, and as an integral part of the course, not with the implication that it is something special for the girls who can't cope with 'real', 'hard' science. Boys as well as girls could benefit from such a reform, which might widen their outlook and increase the status of domestic science in their eyes by associating it with the masculine subject of science. At present the area of science labelled 'domestic' is separated from 'proper' science and taught in a way which renders it more attractive to girls but also gives it lower status and reduces its value as an educational qualification.

This proposal to integrate domestic science into natural science is totally different from the present move to increase the scientific content of domestic science. The current reforms seem likely to confirm the 'ghetto' characteristics of domestic science and girls' exclusion from 'real' science, with its associated power and prestige. By contrast, introducing domestic science into natural science would use girls' existing interest in domestic topics as a springboard to develop wider interests in science. The proposal is also distinct from the suggestion in the Newsom Report (1963, p. 145) that 'for some girls a course might be based on investigating the nature and qualities of apparatus and materials used in the home'. Newsom seems to envisage a separate course of feminine science for girls, rather than introducing feminine topics into the general science course, and such segregation will inevitably lead to second-class status. Some of his practical suggestions as to how topics might be taught so that they appeal to girls and build upon their existing interests and ideas are valuable, but his work contains little suggestion that these interests should be broadened or the ideas challenged.

A more immediately feasible reform, which could be introduced in any school, would be to reduce the amount of freedom pupils have to drop science. At present the crucial choices between subjects are made around the age of thirteen or fourteen. Pupils who drop science at this time are unlikely to either want or be able to take it up again later on. Yet at this age, around the time of puberty, girls are centrally concerned with establishing and defining their femininity (see Chapter 4). They will frequently avoid any activity with masculine connotations.[2] Teenage girls may be totally absorbed with boys, but this does not mean that they want to be 'one of the boys'; it is more likely to lead to a strict demarcation of roles. If they are presented with options at this time girls will tend to act out their femininity by avoiding 'boys' subjects' such as science and choosing 'girls' subjects' such as languages instead. If we really want to encourage girls in science we should seriously consider Milton Ormerod's suggestion (Chapter 7) that these initial choices should be postponed to a later stage.

This has already been tried in some schools, with impressive results. For example, at Hemsworth High School, a thirteen-to-eighteen comprehensive in Yorkshire, eight periods of science per week have been made compulsory for all pupils up to school leaving age. Four periods per week are allocated to biological science and four to physical science, and the courses lead to O levels or CSE's in physical science, biology, applied science, rural science or human biology, depending on academic level. The effect on girls has been dramatic. Prior to the change only 10 per cent of A level physics candidates were female, but now the figure is 40 per cent. Many of these girls say they would have dropped physical science at an earlier stage if they had been allowed to. Postponing the choice also gave many school leavers and non-specialists a more thorough grounding in science than they would otherwise have had. It meant that choices between arts and science were not made until the sixth form, when pupils were more secure in their gender identity and able to make their choices on more rational grounds than might have applied in the fourth form. In addition all pupils experienced a balanced curriculum, including a substantial but not disproportionate science component, until the age of sixteen.

Some schools may consider such changes impossible because of a shortage of science staff. This is often a chicken-and-egg problem, in that staff and facilities are provided only in response to proven demand, but it is difficult to demonstrate the demand without the facilities. A courageous policy of pressing ahead with changes despite inadequate provision and then demanding what is clearly necessary is probably most effective in the long run—although it may be traumatic in the short term.

When it comes to appointing staff, Heads might do well to consider the balance of men and women in the science department. Although there is little real evidence on this, it has frequently been suggested that more women science teachers will have a positive effect on girls' attitudes to science. Certainly an exclusively male science staff will do little to dispel the idea that science is a masculine subject, however personally sympathetic the teachers may be to girls in science. Equally, if the women science teachers are concentrated in biological science the stereotypes will be reinforced. Women teachers of physical science may act as role models for the girls, living proof that women can succeed in this area. They may also, because of their personal experience, be more sympathetic than men teachers to girls' learning styles and interests. Galton (Chapter 13) suggests that women teachers are more likely to employ the teaching styles which girls prefer, but unfortunately the number of women teachers in his study is small, particularly in physical science, so it is difficult to draw any definite conclusions. Of course, sex should not be allowed to override qualifications, experience or individual personality in determining appointments, but, other things being equal, it is probably desirable that there should be more women physics and chemistry teachers.

It was suggested in the previous section that teachers should be aware of, and make allowances for, possible gaps in girls' scientific backgrounds. A more formal approach to the same problem would be to set up remedial classes in schools. Children would be encouraged to play with apparatus and explore its properties in a more leisurely atmosphere than exists in the science lessons. Educational toys would be made available, although these should not be too obviously 'boys' toys' which the girls might find off-putting. Such classes would be available to children of either sex who needed them, although they would probably be used mainly by girls (in the way that remedial reading classes are used mainly by boys). However, they should probably not be withdrawal classes, since science is a cumulative subject and withdrawal might only widen the gap between high and low achievers; instead classes could be run in lunch hours or after school, or timetabled against optional 'interest' subjects. The aim would be to compensate for any deficiencies in girls' science-related knowledge or spatial skills, and enable them to begin the formal study of science with the same reserves of incidental knowledge as the boys. The growth of science and technology in primary schools should also be encouraged, since this may have a similar effect if it enables girls to gain confidence in manipulating apparatus and finding out how things work.

Remedial classes are already availabe in a variety of contexts. Tobias (1978) describes the Maths Anxiety Clinics which are becoming increasingly common in the United States. These are intended to help female students overcome their fear of the subject and build up their confidence by going back to elementary arithmetic and building mathematical competence on a solid foundation of understanding of basic operations. Although they usually operate at college level, these clinics are similar to school-level remedial science classes in that they aim to fill gaps in basic understanding before the pupils move on to more advanced ideas. In Britain there are a few science and technology courses in further education and the Open University intended for students (particularly women) with no background in these subjects. (Levy, 1978; Swarbrick, 1979). At Holly Lodge, a comprehensive girls' school in Liverpool, the staff, in conjunction with the local science advisers and employers, have devised a Mode 3 CSE course in basic electronics which is specifically designed to help pupils with little or no background knowledge to prepare themselves for employment or training in technical subjects. The course involves a large component of practical and project work, and after some initial hesitation the girls seem to be responding enthusiastically (Hannon, 1979; Mainwaring, 1979).

Although it is not remedial, girls may also benefit from another type of special class. In a survey of schools which were particularly successful in encouraging girls in science, HM Inspectors found that several of them had instituted work clinics (Harris, 1979). Pupils could come along after school to sort out any problems they might have had in the science lessons. It was found that these clinics were used more by girls than boys, and that very often the girls did not really have a problem, but were just seeking reassurance that they had understood the material correctly. The girls seemed to be unhappy if they were insecure in their work, and the clinics were important in allaying their fears.

School policies

Many of the changes which might benefit girls in science would affect the whole school and not just the science department. As such they are dependent upon overall school policy. It is difficult to make specific suggestions here, because internal school organisation varies so widely, but some general points are applicable to many situations.

For example, timetabling may have the unintended consequence of dissuading girls from science if the physical sciences are set against subjects which are traditional favourites with girls. Timetables should be examined with the following questions in mind. Is it easier to take physical science with technical drawing and metalwork or with domestic science and needlework? With geography or with history? With economics or with modern languages? Do our notions of what subjects 'go well together' owe most to tradition, to vocational requirements or to considerations of educational balance? Are subjects such as human biology which are popular with girls only offered at CSE and not O level? The answers would probably tell us much about our unconscious biases.

Careers advice can also be influential in encouraging girls in science. The crucial nature of the choices between subjects at the end of the third year is a theme which has recurred throughout this book. Yet in many schools pupils and their parents are left to make these choices with little guidance from anyone who is conversant with the long-term implications of their decisions. In particular, careers guidance is frequently provided for fifth formers but not for third formers, despite the fact that choices made at the end of the third year may effectively rule out a large number of careers in the future. But as the pupils' letters in Chapter 17 show, when girls are carefully counselled they will take science subjects for career reasons even if they dislike or are having difficulty with the subjects. Several of these girls showed an acute awareness of the fact that they would have a wider choice of careers if they continued to study two or more science subjects, and of the relevance of science to traditional female jobs such as hairdressing and nursing. Hearn (1979) describes the extensive programme of talks and exhibitions provided by his school (Bedford Park Comprehensive, Romford) for third year pupils and their parents to persuade them of the importance of physical science for girls. Careers, guidance and science staff were all involved and class discussions were also employed. The importance of physical science for traditional female jobs involving biology (such as nursing and hairdressing) was stressed, so as to build upon the girls' existing interests. The increase in the proportion of girls studying physical science, particularly chemistry, since the introduction of this programme has been quite dramatic.

Most girls do not have much contact with women scientists, and many girls need active encouragement to consider science-based careers and to understand that these can be compatible with a feminine identity. Contact with women working in a variety of scientific jobs at all educational levels may help to dispel some schoolgirl myths. In the United States the National Science Foundation has funded a variety of projects based on this idea (Fox, 1976b). One of the most successful attempted to provide female role models for jobs in science by arranging for groups of women scientists and technologists to visit schools to talk about themselves and their jobs (Weiss *et al.*, 1978). A pack of careers material was also supplied and the girls were encouraged to write away for more information. If live models are not available it may be possible to use books or films of women scientists, though preliminary results with this method are not so encouraging (Fox, 1976b).

But careers advice is meaningless without job opportunities. It is all very well to encourage girls in school to broaden their horizons and to consider craft and technical occupations. Many teachers are understandably unhappy about doing this if they feel that the employers in their area are unlikely to accept girls for training in these jobs. The schools and the Careers Service should therefore make it their business to develop links with industry which will provide openings for girls. The advantages for the employer of selecting from a wider field of candidates easily outweigh any practical considerations such as the provision of toilet facilities (more often an excuse than a real problem—how many firms have no female typists, tea ladies or cleaners?). And as Peggy Newton has noted (Chapter 10) when one girl from a school has shown the way into an unconventional occupation others frequently follow.

The idea that women scientists are in some way strange or abnormal could also be countered by greater awareness of the part women have played in science. With the exception of Marie Curie—the exception that proves the rule—women's contributions are frequently ignored or underrated. (Try asking your colleagues how many women scientists they can name!) But as Mozans (1974) has shown, throughout history women have made important scientific discoveries. Greater awareness of this among girls (and their teachers) might do much to boost the girls' self-confidence. Class discussion of women's achievements and the reasons for their under-achievement in various spheres could usefully be incorporated into social science or history lessons as well as science lessons.

As Judith Whyte (Chapter 19) points out, sex stereotyping in science is an aspect of more general sex stereotyping in schools, and should be tackled as such. Hannon (1979) suggests a variety of measures for combating it, prominent among which are school-based workshops. These could involve discussions, video-taping and role-playing designed to heighten teachers' awareness of the ways in which they differentiate between the sexes. Similar workshops in teacher-training and in-service courses would also be valuable. Non-sexist teaching advisers are also a possibility, and if the problem were considered sufficiently important each school could designate a staff member with special responsibility for countering sexism. As Hearn (1979) says, 'the most difficult stage in the whole exercise is accepting the problem and resolving to do something about it'. The specialist's task would be to encourage cross-sex activities throughout the school, to point out to teachers what aspects of their behaviour might be discriminatory and to suggest alternative ways of coping with the situation. However, it should always be made clear that the specialist is only an adviser, and that reducing sexism in school is everyone's responsibility.

In the last resort, some problems can be tackled with the aid of the law. Byrne (1975) found that girls' schools were more likely than mixed or boys' schools to have too few science laboratories for the number of pupils. She also found that most of the mixed schools 'timetabled boys into science laboratories for chemistry and physics, teaching biology to girls in converted classrooms, simplistic labs with missing facilities or not at all'. The Sex Discrimination Act has made such blatant discrimination illegal, but giving girls and boys the same formal choice does not guarantee equality. There may be informal norms that define which subjects are for girls and which are for boys, and it takes a very determined child (or parent) to break them. As Peggy Newton reports (Chapter 10), many co-educational schools continue to offer some subjects, particularly crafts, to one sex only. Technical drawing is clean and light, requiring neatness and manual dexterity, but that does not prevent it being a 'boys' subject' and it may take the threat of a court case to get any alteration. Strathclyde has recently changed its policy on this issue, following pressure from the Equal Opportunities Commission (EOC, 1979). However, the Act does not apply to single-sex schools, and so cannot affect discrepancies between different types of school. The justification for this is largely pragmatic—it would undoubtedly cost a lot of money to ensure that girls' schools were as well equipped for science as boys' schools—but it seems morally indefensible.

The differences between single-sex and co-educational schools crop up regularly in any discussion of girls and science. As already mentioned, the facilities and the staffing for science are frequently better in co-educational schools than in equivalent girls' schools. Girls may also have a wider choice of subjects, particularly the traditional 'boys' crafts in co-educational schools. Nevertheless, the burden of the evidence is that girls are more likely to choose physical science and more likely to do well in it if they do choose it in single-sex schools than in mixed ones (see Chapters 7, 12 and 14; see also

DES, 1975). The most likely explanation for this pattern is that girls are disadvantaged in a mixed school because science is dominated by boys and is clearly labelled as a boys' subject in a way that cannot occur in a girls-only school. Alternatively, girls may be disadvantaged if teachers react to the greater enthusiasm and knowledge of the boys by subconsciously gearing their teaching to the boys' interests and learning styles.

Nevertheless, girls do succeed in science in some mixed schools, and a variety of strategies for overcoming their problems have already been suggested. Another approach is currently being investigated at Stamford Secondary School, Ashton under Lyne. The staff there were worried by the low proportion of girls opting for physical science in the fourth form, and impressed by the evidence that girls do better at science in single-sex schools. They therefore decided to teach mathematics and science to some of their first and second year pupils in single-sex groupings and see how this affected girls' attitudes and achievements; the remainder were taught in mixed classes as usual, and formed a control group. The experiment has only been running for a year, and the school sensibly decided not to carry out a full evaluation until the end of the two-year period, but first results suggest that the girls' marks have improved. Informal comments indicate that the experiment is popular among the girls, who like being able to get on with their work without being disrupted by the boys; but the teachers have reacted less favourably, and say they do not enjoy teaching the girls-only classes.

Wider issues

Schools and school science do not exist in a vacuum, but should be located in a wider social context. The suggestions for intervention in this chapter have concentrated on school-based strategies not because school is necessarily the most important influence on girls' choice of science but because it is the easiest to change. By the time they enter secondary schools, girls (and boys) are already well socialised into sex roles, and they already have a picture of science. When they leave they enter a social and occupational world which is riddled with sexual stereotypes and prejudices. Intervention is also important in these areas, though it may take time for it to be effective. Strategies for change can be roughly divided into two broad groups: those concerned with changing girls and those concerned with changing science.

Strategies to change girls revolve around reducing sex stereotypes and expanding the definition of femininity beyond the bounds of 'Kinde, Küche and Kirche', children, the kitchen and the church. Such strategies must operate through the home, the media and the primary school as well as through the secondary school. Boys are often brought up to be more independent than girls, and independence training seems to be linked to the qualities necessary for a successful scientist (Maccoby, 1970). The female role is usually depicted by the media and children's reading books as solely that of wife and mother despite the fact that over 60 per cent of adult women in Britain are in paid employment. Pressure groups and teacher training courses both have their part to play in countering these influences by attacking stereotyped images and making the reality of women's employment more widely known. If girls realise that they will probably work outside the home for thirty years of their life they may take the question of training for a career more seriously. Society's conception of femininity must be broadened to include a combination of home and career commitments.

There are in fact some signs of change in accepted notions of femininity; but there is little sign of the corresponding changes in the notion of masculinity which are necessary for women to develop their careers fully. Men have to accept an equal part in housework and child care if women are to take an equal part in employment. Employers have to expect and accept that their employees have home commitments and may require intermittent or part-time employment. As husbands, fathers and employers men have something to lose by these changes (although also something to gain), and only sustained pressure from women will bring them about.

Strategies to change science are more frequently school-based, although they too extend into the wider society. Two of the principle characteristics of physical science are its impersonality and its masculine image. I do not believe that either of these qualities is intrinsic, although both are probably off-putting to girls. Neither is so prevalent in biology, which may account for the greater popularity of biological science with girls. Ways in which the impersonality and masculinity of science might be combated in school have already been discussed, but their wider consequences are disquieting for all concerned.

It is sometimes suggested that modifications to science teaching designed to encourage girls will be disadvantageous to boys. Personally I consider this extremely unlikely (although even if it were true I would defend steps to distribute advantages and disadvantages equally to girls and boys). The measures to humanise and feminise science which have been proposed in this chapter should benefit both sexes. The benefits to girls are in terms of increased enrolments on science courses; the benefits to boys are in terms of a more sensitive understanding of the place of science and of women in society.

Science students today, both boys and girls, but especially the boys, score far lower than arts students on scales of person orientation (Chapter 12). Working scientists are also typically uninterested in people (Chapter 16). Science seems to attract this type of person by its apparent impersonality. But the implications for society of anything as powerful as science being controlled by men with little or no interest in, or understanding of, social relations are frightening. It is not sufficient to encourage more girls to become substitute boy scientists, even if that were possible. If school science were reformed to make it more humanistic we might well find a different sort of person (and probably a higher proportion of women) being attracted into it, and a different sort of science being produced. This in turn might improve the discipline's tarnished reputation in the world.

Because science is generally considered a 'boys' subject', girls who venture to study it often find themselves in a hostile male environment (see the pupils' letters, Chapter 17). To persevere, they have to be very determined and unhindered by a concern for social approval. Smithers and Collings (Chapter 12) found that girl scientists have a negative opinion of their own social attractiveness. They suggest that science may be a haven for girls who have difficulties with social relations. But it seems equally likely that society is exacting a price from those who violate its norms by convincing them that they are social rejects. If science is a haven, why do boys with similar problems not seek it out? Girls in science are continually told, 'That's a funny thing for a girl to study,' and they may come to believe it and see themselves as social misfits. The masculine image of science probably causes considerable frustration and waste of talent. As Bradley (Chapter 11) points out, many more girls than currently study science have the intellectual capacity to do so. If sex stereotypes play a major part in determining children's education, then a large number of girls and boys will be forced into roles which do not fit their individual aptitudes and interests.

However, some authors have argued that far more than sex stereotyping is involved in the masculine image of science. They suggest that science, and the ways of thinking and acting associated with it, with their connotations of control and manipulation, are intrinsically masculine (Stéhelin, 1976; Wallsgrove, 1980). Others have suggested that women are deliberately excluded from science as part of their exclusion from positions of power in society (Saraga and Griffiths, Chapter 6). Neither of these positions seems convincing to me. I do not believe that women and men are as different as is implied by the first argument, nor that patriarchy is as well organised as is implied by the second. Rather, I would suggest that science is masculine because it is dominated by men. The things men do acquire masculine connotations and these connotations are important in ensuring that men continue to do them. The way in which science came to be dominated by men has not been properly researched. But it may be possible to reverse the process by attacking the masculine connotations of science and so rendering it more attractive to women.

The reforms necessary to encourage girls in science will not be easy. As Saraga and Griffiths suggest (Chapter 6), they run counter to many vested interests. A political will is required if women's position in society is to change. But changing women's position in science is part and parcel of changing women's position in society, and the struggle can take place on this front as on others. To take the attitude that nothing can be done about girls in science until women's position in society has been altered is both defeatist and likely to lead to defeat. Women's position in society will not improve unless a concerted effort is made to improve it in all walks of life, including science.[3]

Notes

1 In contrast to this approach, Saraga and Griffiths (Chapter 6) argue that the present concentration on industrial and military aspects of science is intrinsically bound up with the development of the subject.

2 Conversely, and possibly even more rigorously, boys will avoid anything with overtones of femininity.

3 I would like to thank Judy Samuel and Elinor Kelly for reading and commenting on an earlier draft of this chapter.

References

Byrne, E. M. (1975), 'Inequality in education: discriminal resource allocation in schools', *Educational Review*, Vol. 27, 179.

Children's Rights Workshop (1976), *Sexism in Children's Books: Facts, Figures and Guidelines*, Writers' and Readers' Publishing Co-operative.

Department of Education and Science (1975), *Curricular Differences for Boys and Girls*, Education Survey 21, HMSO.

Equal Opportunities Commission (1979), *EOC Victory for Strathclyde Schoolgirls*, press notice, 27 June.

Federbush, M. (1974), 'The sex problems of school maths books' in Stacey, J., Béreaud, S., and Daniels, J. (eds.), *And Jill Came Tumbling After*, Dell, New York.

Fox, L. H. (1976a), 'Sex differences in mathematical precocity: bridging the gap', in Keating, D. P. (ed.), *Intellectual Talent: Research and Development*, Johns Hopkins University Press.

—(1976b), 'The effect of sex-role socialization on mathematics participation and achievement', Education and Work Group, National Institute of Education, USA, December.

Graf, R. G., and Riddell, J. C. (1972), 'Sex differences in problem solving as a function of problem context', *J. Educational Research*, Vol. 65, 451.

Hannon, V. (1979), *Ending Sex-Stereotyping in Schools: A Sourcebook for School-Based Teacher Workshops*, Equal Opportunities Commission.

Harris, B. (1979), Paper presented at 'Girls and Science' course, Wooley Hall College, West Yorkshire, 14 June.

Hearn, M. (1979), 'Girls for physical science: a school-based strategy for encouraging girls to opt for the physical sciences', *Education in Science*, April.

Hutchings, D. (1967), 'Girls' attitudes to science', *New Society*, 9 November.

Ingle, R. B., and Shayer, M. (1971), 'Conceptual demands in Nuffield O level chemistry', *Education in Chemistry*, Vol. 8, 182.

Levy, M. (1978), 'Part-time science—"more with women in mind" ', *South London Press*, 22 September.

Maccoby, E. E. (1970), 'Feminine intellect and the demands of science', *Impact of Science on Society*, Vol. 20, 13.

Mainwaring, E. E. (1979), 'A school course in basic electronics', *International Journal of Electrical Engineering Education* (in press).

Milton, G. A. (1958), 'Five studies of the relation between sex role identification and achievement in problem solving', Yale University Technical Report, 3.

Mozans, H. J. (1913, rpr. 1974), *Woman in Science*, MIT Press.

Newsom Report (1963), *Half our Future*, HMSO.

Rowell, J. A. (1971), 'Sex differences in achievement in science and the expectations of teachers', *Australian J. of Education*, Vol. 15, 16.

Rutherford, F. J., Holton, G., and Watson, F. G. (1975), *Project Physics*, Holt Rinehart & Winston.

Shayer, M. (1972), 'Conceptual demands in the Nuffield O level physics course', *School Science Review*, Vol. 53, 26.

Smail, B., Kelly, A. and Whyte, J. (1980), *Girls Into Science and Technology*, Didsbury Faculty, Manchester Polytechnic.

Stéhelin, L. (1976), 'Science, women and ideology', in Rose, H. and S. (eds.), *The Radicalization of Science*, Macmillan.

Swarbrick, A. (1979), 'Women and technology in the O.U.', *Women and Education Newsletter*, No. 16.

Tobias, S. (1978), *Overcoming Math Anxiety*, Norton.

Wallsgrove, R. (1980), 'Towards a radical femininist philosophy of science', in Brighton Women and Science Collective (eds.), *Alice through the Microscope*, Virago.

Weiss, I. R., Pace, C., and Conaway, L. E. (1978), *The Visiting Women Scientists Pilot Program 1978 Final Report*, Research Triangle Institute, North Carolina 27709, USA.

Statistical Appendix

ALISON KELLY

Some of the research-based chapters in this book necessarily make use of statistical terms. In order to increase the accessibility of these chapters to the general reader, this Appendix provides a brief guide to the use and meaning of these terms. The derivation and mathematical properties of the statistics are, however, beyond the scope of this discussion. For a more technical account the reader is referred to a standard text (e.g. Blalock, 1960; Guildford, 1965).

When we speak of the average we are usually referring to the *mean*. For example, if the science scores of eleven children are 6, 7, 7, 7, 8, 8, 9, 10, 10, 11 and 21 the mean will be

$$\frac{6+7+7+7+8+8+9+10+10+11+21}{11} = 9{\cdot}5$$

If the scores are respresented by X_i and the number of cases by N, then the mean is written $\overline{X}$ and is defined as

$$\overline{X} = \frac{\Sigma X_i}{N}$$

There are in fact two other kinds of average, the median and the mode, which are sometimes used. The *median* is the middle score (8 in the above example, with five children scoring above and five scoring below this value). The *mode* is the most common score (7 in the above example, scored by three children). The mean is the most statistically convenient average and is particularly useful when the scores are normally distributed. A *normal distribution* has an exact mathematical form. In general terms, it is bell-shaped, with most of the scores bunched in the middle, only a few high and low values, and no particularly deviant scores (see Fig. A1). In the example 21 is a deviant score, being very much above the others. The mean is

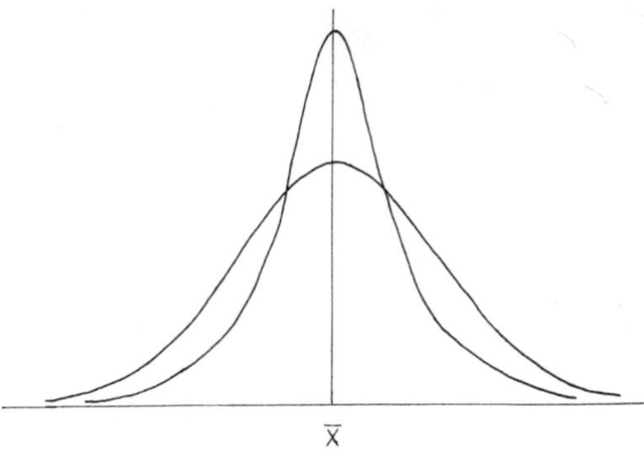

Fig. A.1 Two normal distributions with the same mean but different standard deviations.

substantially increased by this one aberrant value. Perhaps a better indication of a typical value is provided by the median (8). In the example, only a few children have each score and it is not very important that three score 7 while only two score 8 or 10, but in some cases the most typical value is the most common value. For a normal distribution the mean, mode and median coincide—and normal distributions are, as their name implies, the most common type, particularly for naturally occurring characteristics. Several of the statistics described below (for example, correlations and *t* tests) are only strictly applicable to normal distributions.

The *standard deviation* (s.d.) is a measure of how spread out a set of scores is. If everybody scores the same, the standard deviation will be zero. If the scores are all similar the s.d. will be small, but if they vary a lot, the s.d. will be large (see Fig. A.1). To make the idea of spread more precise we can consider the average deviation of the scores from the mean. But a simple average turns out to be zero; because of the way the mean is defined, the high values are as far above the mean, overall, as the low values are below the mean. In order to get a non-zero value for the average deviation we have to treat all deviations as positive. In calculating the standard deviation this is done by squaring. This, of course, gives us 'square units', so we take the square root. This yields a measure of how much the scores vary about the mean. In mathematical terms the standard deviation is defined as

$$\sqrt{\frac{\Sigma (X_i - X)^2}{N}}$$

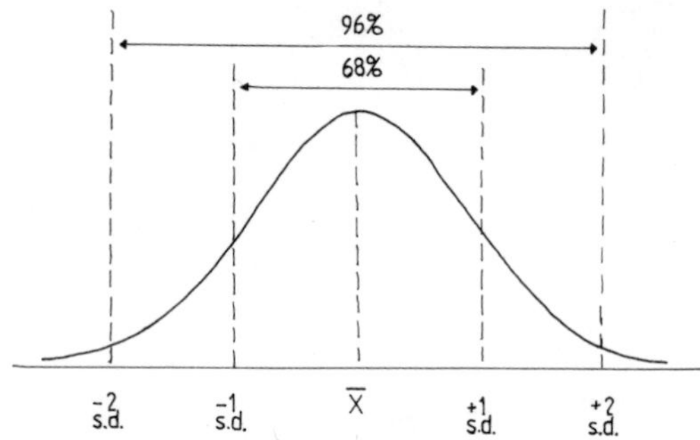

Fig. A.2 The proportions of a normally distributed sample lying within one and two standard deviations of the mean.

When a set of scores is normally distributed the variation about the mean is precisely known in terms of the standard deviation. Sixty-eight per cent of the population fall within 1 s.d. of the mean and 96 per cent fall within 2 s.d. of the mean (see Fig. A.2). For example, with IQ scores standardised to a mean of 100 and an s.d. of 15, 68 per cent of the population have scores between 85 and 115, and 96 per cent have scores between 70 and 130.

When comparing the standard deviations of different distributions we must, of course, be careful to compare only variables which are measured on the same scale. It is valid to compare the s.d. of the scores of girls and boys on the same test. But it is not valid to compare the s.d. of girls' scores on two different tests, since this will depend on the way the test has been constructed. It is, however, valid to compare proportions of a standard deviation for variables measured on different scales—this is the point of the standardisation. Thus it is permissible to say that boys are, on average, 61 per cent of a standard deviation ahead of girls on the IEA physics test, but that the gap on the biology test is only 13 per cent of a standard deviation (see Chapter 2).

When, as is nearly always the case in social science research, we are measuring only a sample of the people we are interested in, and not the whole group, the values that we obtain for the mean and standard deviation are only estimates of the mean and standard deviation of the whole group. In statistical tests, the whole group is usually called the *population*, even when it is only referring to a limited group of people such as sixth-form science specialists. If we had drawn another sample of the same size we would have got slightly different estimates. The accuracy of the estimates depends on the size of the sample—the larger the sample the more likely are our estimates to correspond to the true values for the whole group. The *standard error* of the mean indicates the probable accuracy of our estimate of the mean. The standard error is equivalent to the standard deviation of the distribution of means we would get if we drew repeated samples of the same size. 68 per cent of the time the mean that we get from a sample will be within one standard error of the true mean for the whole group, and 96 per cent of the time it will be within two standard errors of the true mean.

Often we want to know how likely it is that the result we have found has arisen by chance because of the way we have chosen our sample, when there is no such pattern in the whole population. In other words we want to know how certain we may be that we have found a real effect, and not just a statistical fluke. To determine this, we must perform a *test of significance*. We set up the *null hypothesis*, that in the population we are studying there is, in reality, no difference between the groups we are studying (e.g. no difference in the science scores of girls and boys). If we have in fact observed a difference between the scores of the girls and boys we tested we want to know whether this has just occurred by chance because we happened to pick a particularly high-scoring group of boys and/or a particularly low-scoring group of girls when drawing our sample (assuming that the sample has been correctly drawn—significance tests do not allow for any bias in the sampling procedure). We can, of course, never be sure that the observed difference has not arisen by chance, but we can calculate the probability that it is a chance result. Conventionally we accept that a result is probably real if it will occur by chance, because of the way the sample has been chosen, less than one time in twenty. This is called the 5 per cent level of significance—there is a 5 per cent chance that the pattern is *not* present in the whole population. More stringent tests demand a 1 per cent level of significance (one time in a hundred the result will arise by chance) or a 0·1 per cent level (one time in a thousand the result will arise by chance).

It is worth noting that statistical significance does not necessarily correspond to educational or social significance. With large samples even a small difference between groups will often turn out to be statistically significant. This does not mean that it is an important difference, only that it is unlikely to have arisen by chance. There are no statistical criteria that can tell us what differences between groups are important. That is purely an educational or social judgement.

There are a large number of different significance tests, appropriate to different situations. One of the most common is Student's *t test*. This is used for comparing the means of two sets of scores. The test is more likely to be significant if there is a large gap between the means, if the standard deviations are small and if there are large numbers of people in each group. To compare the means of more than two groups the *F test* associated with analysis of variance (see below) has to be used. If the variables are such that it is not meaningful to compute an average then the *chi squared* (χ^2) *test* is commonly used, For example, if we are interested in comparing the proportion of girls and boys naming parents, teachers and peers as the major influence on their subject choice, the χ^2 test can be used to see whether proportions of responses in each category are significantly different for girls and boys. χ^2 can also be used to compare proportions among several groups of people (e.g. arts, science and social science specialists) simultaneously. The *Mann–Whitney U test* is used to compare two sets of rankings. For example, if a group of schoolchildren are asked to rank their school subjects in order of preference the Mann–Whitney U test can be used to compare the rankings given to any one subject, such as physics, by girls and boys.

When the appropriate test has been chosen, the statistics have to be computed according to formulae. Distribution tables (which are reprinted in most statistical textbooks) are then consulted with the appropriate sample size (N) or *degrees of freedom* (*d.f.*) to determine their level of significance. The degrees of freedom depend not only on the sample size but also on the number of categories into which the sample is divided.

Correlation coefficients are a way of relating two variables both of which should be normally distributed. In practice it is not important that the variables be normally distributed, but it is important that they should both be the sort of variables for which it is meaningful to calculate an average. Thus science scores, scores on attitude scales, teacher's salaries or hours spent doing homework can be correlated; but sex, father's occupation or main influence on subject choice cannot be correlated. Correlation coefficients run from +1·0 through zero to −1·0. A value of +1·0 indicates a perfect positive correlation—as one variable gets larger so does the other, and if one is known the other is exactly known. In practice correlations of +1·0 never occur in social science, but the correlation between, for example, height and shoe size would be fairly large and positive, since tall people usually have larger feet than small people.

Correlations around zero indicate that there is little or no relationship between the two variables being considered. Thus the correlation between height and IQ would probably be near zero, since, to the best of my knowledge, tall people are, on average, no more or less intelligent than small people. Negative correlations occur when one variable gets smaller as the other gets larger. For example, the frequency of thumb sucking among children is probably negatively correlated with height, since bigger children suck their thumbs less than little ones.

In social science, correlations are usually fairly small and anything numerically larger than 0·2 is probably worth reporting. The small size of the correlations indicates that most of the relationships with which we are dealing are complicated and affected by a wide variety of factors. In addition, our measurement techniques are frequently poor, so that the relationships are obscured by large measurement errors.

Correlation coefficients can be interpreted as the average change in one variable for unit change in the other. A correlation of 0·5 indicates that people who differ by one standard deviation on one variable will differ, on average, by 0·5 standard deviations on the other. An alternative interpretation of the coefficients is described under variance, below. Since correlation coefficients are standardised (i.e. interpreted in terms of standard deviations) they can be used to compare extremely diverse relationships. For example, we can examine whether science achievement is more closely related to father's education or to attitude to science, although these two variables are measured in totally different units.

Correlation coefficients are conventionally designated by the letter r. So r=0·5 means a correlation of 0·5. They can only represent straight-line relationships. If one variable first increases and then decreases as the other increases (e.g. number of teeth and age) a correlation coefficient will underestimate the closeness of the relationship.

There are several mistakes which people commonly make in using and interpretating correlation coefficients. As with other statistics, the significance of a correlation coefficient can be calculated in a straightforward way; but with large samples even very small correlations can be significant, and this does not mean that the relationship between the two variables is either large or important. Even more crucially, it must be remembered that correlation does not imply causation. Just because two variables are correlated does not imply that one causes the other. I suggested above that height and frequency of thumb sucking among children might be negatively correlated. This does not mean that thumb sucking causes stunted growth. It is more likely to be a spurious correlation due to the intervening variable, age. Young children generally suck their thumbs more than older children. To check the accuracy of this suggestion we could compute a *partial correlation* between height and thumb sucking, controlling for age. This would have the effect of saying, if we compare children who are all the same age, is height related to thumb sucking? Most likely we would find that the partial correlation was near zero. But even if it were still substantial, this would not prove that thumb sucking produces stunted growth. Correlations can never prove causation (although they can strengthen a causal argument made on other lines). If we seriously wanted to suggest that thumb sucking produces stunted growth we would have to provide some sort of physiological explanation as to how this effect might occur as well as the correlational evidence.

Partial correlation can be used to control for more than one variable at the same time. Thus if we wanted to pursue the thumb-sucking problem we might compute the partial correlation between height and thumb sucking controlling for babyish appearance and parental strictness as well as age. Ordinary correlations are sometimes referred to as *zero-order* correlations in order to distinguish them from partial correlations.

Variance is a useful statistic which crops up in many situations. It is the square of the standard deviation and is defined as the average of the squared deviations from the mean

$$\frac{\Sigma(X_i - X)^2}{N}$$

As such it is a measure of how much variation there is in a sample, and it is when considering variation that the variance becomes important.

When analysing data we are often concerned to see how well we can predict people's scores on one variable such as science achievement (the *dependent variable*) if we know their scores on other variables such as sex, father's occupation or attitude to science (the *independent variables*). Generally, we cannot predict the dependent variable completely accurately, but we can make a better guess at it if we know the scores on the independent variables than if we do not know these scores. People with the same scores on the independent variables will frequently vary less among themselves on the

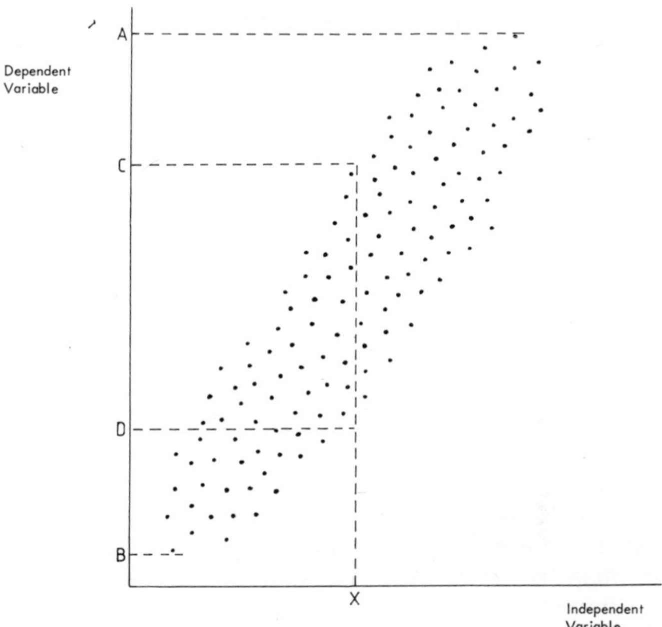

Fig. A.3 The reduction in variation in an independent variable when the value of a dependent variable is known.

dependent variable than will all the people in the sample. Our guess as to the value of the dependent variable is improved to the extent that we have reduced the variation in this variable. This is illustrated in Fig. A.3, which shows two variables with a fairly strong correlation. If we have no information about the independent variable, we know only that the dependent variable falls within the range A to B. But, if we know that the value of the independent variable is X then we know that the value of the dependent variable falls within the restricted range C to D. The variation is reduced from AB to CD. This is often expressed as the independent variable having 'explained' some of the variation in the dependent variable. Clearly it is only explained in a statistical sense, not in any theoretical way, since we may still not know why the variables are related as they are.

This idea of reducing or explaining the variation is quantified by the use of the variance statistic. The total variance can be divided into the explained and the unexplained. This gives us another interpretation of the correlation coefficient. The square of the correlation coefficient is the proportion of the variance in one variable explained by the other variable. So with a correlation of 0·5 we can say that $0{\cdot}5^2 = 0{\cdot}25$ or 25 per cent of the variance is explained. The other 75 per cent is still unaccounted for. In other words the variance of the scores on the dependent variable for people with the same score on the independent variable will average out at 75 per cent of the total variance. Since the variance is the square of the standard deviation, the standard deviation of the scores on the dependent variable for people with the same score on the independent variable will be $\sqrt{0{\cdot}75} = 0{\cdot}87$ or 87 per cent of the standard deviation of all the people in the sample. Note that quite a respectable correlation of 0·5 produces only a 13 per cent reduction in the standard deviation. If the correlation is smaller, for example, $r=0{\cdot}2$, then only 4 per cent of the variance is explained, which means that 96 per cent of it is unexplained, and the standard deviation is only reduced by 2 per cent ($\sqrt{0{\cdot}96}=0{\cdot}98$).

This technique of dividing up the variance into explained and unexplained portions is not confined to the interpretation of correlation coefficients. It is widely used in data analysis, particularly in *analysis of variance*. In its simplest form analysis of variance is used to compare the means of a dependent variable in several categories of an independent variable (for example, the science scores of children from different fathers' occupational categories). It yields the statistic F, and Tables of the distribution of F have to be consulted to determine whether the group means differ significantly. As usual, the significance of F depends not only on the means and standard deviations in the various categories but also on the size of the sample. More complicated analysis of variance can be used to compare several sets of categories simultaneously (for example, the science scores of children of different sexes in different fathers' occupational categories) and to check for *interactions* between the categories (i.e. in this example to check whether the gap between the sexes differs significantly in the different fathers' occupational categories).

A brief appendix such as this cannot cover all the statistics employed in social science. Nor is it possible to give an adequate description of the more complicated techniques without at least touching on their derivation and definition. Several of the more advanced procedures, such as *regression*, *path analysis* and *discriminant analysis* are concerned with the prediction of a dependent variable from a number of independent variables. They differ in the mathematical procedures used to make the prediction and in the extent to which theory or statistical convenience determines the way in which the variables are used. In discriminant analysis a series of predictor variables are chosen and combined in such a way as to make the groups as statistically different as possible. Scores on a science test, attitudes to science and personality variables might all be suitable predictors for distinguishing between pupils who, at a later stage, would become arts or science specialists (see Chapter 11). In regression or path analysis, these variables would be entered in a theoretically determined order, possibly corresponding to the stage of a person's life at which they operated. But in discriminant analysis the order is determined solely by the degree to which the variables distinguish between the groups.

Other techniques, such as *factor analysis* and *cluster analysis* are used to identify groups of variables (factor) or individuals (cluster) which go together. Factor analysis is most commonly used in scale construction. For example, if we are interested in measuring attitudes to science we cannot just ask pupils, 'What is your attitude to science?' Instead we ask them a large number of more concrete questions such as 'Do you enjoy science lessons?' or 'Would you like to be a scientist when you leave school?' Factor analysis is then used to pick out the underlying patterns in pupils' responses, and identify the questions which seem to be tapping the same dimension. Cluster analysis picks out the individuals who, taking their responses to a number of questions together, seem to be most similar in their opinions or characteristics.

Data not only have to be collected and analysed, they also have to be interpreted, and every stage has its dangers. As Jencks *et al.* (1972, p. 358) have pointed out, the same information can be presented in a variety of ways so as to maximise or minimise its impact. There are no hard-and-fast rules about how data should be presented, and even fewer about how it should be interpreted. Authors are the people most familiar with their data, and they should guide their readers towards an interpretation. But readers should be aware that they are being guided and that other interpretations are possible. Anyone who intends to make a habit of reading research reports would be well advised to spend an evening with Darrell Huff's *How to Lie with Statistics* (1954), an invaluable aid in detecting the statistical sleight of hand!

References

Blalock, H. M. (1960), *Social Statistics*, McGraw-Hill.
Guildford, J. P. (1965) *Fundamental Statistics in Psychology and Education*, McGraw-Hill.
Huff, D. (1954), *How to Lie with Statistics*,
Jencks, C., *et al.* (1972), *Inequality*, Basic Books.

NAME INDEX

SUBJECT INDEX